Lecture Notes in Economics and Mathematical Systems   584

Founding Editors:

M. Beckmann
H.P. Künzi

Managing Editors:

Prof. Dr. G. Fandel
Fachbereich Wirtschaftswissenschaften
Fernuniversität Hagen
Feithstr. 140/AVZ II, 58084 Hagen, Germany

Prof. Dr. W. Trockel
Institut für Mathematische Wirtschaftsforschung (IMW)
Universität Bielefeld
Universitätsstr. 25, 33615 Bielefeld, Germany

Editorial Board:

A. Basile, A. Drexl, H. Dawid, K. Inderfurth, W. Kürsten, U. Schittko

T0190428

Charlotte Bruun (Ed.)

Advances in Artificial Economics

The Economy as a Complex
Dynamic System

With 93 Figures
and 30 Tables

 Springer

Editor

Professor Charlotte Bruun
Department of Economics, Politics
and Public Administration
Aalborg University
9100 Aalborg, Denmark
cbruun@socsci.aau.dk

ISBN-10 3-540-37247-4   Springer Berlin Heidelberg New York
ISBN-13 978-3-540-37247-9   Springer Berlin Heidelberg New York

This work is subject to copyright. All rights are reserved, whether the whole or part of
the material is concerned, specifically the rights of translation, reprinting, reuse of
illustrations, recitation, broadcasting, reproduction on microfilm or in any other way,
and storage in data banks. Duplication of this publication or parts thereof is permitted
only under the provisions of the German Copyright Law of September 9, 1965, in its
current version, and permission for use must always be obtained from Springer-Verlag.
Violations are liable for prosecution under the German Copyright Law.

Springer is a part of Springer Science+Business Media
springeronline.com

© Springer-Verlag Berlin Heidelberg 2006

The use of general descriptive names, registered names, trademarks, etc. in this publication does not imply, even in the absence of a specific statement, that such names are
exempt from the relevant protective laws and regulations and therefore free for general
use.

Typesetting: Camera ready by author
Cover: Erich Kirchner, Heidelberg
Production: LE-TEX, Jelonek, Schmidt & Vöckler GbR, Leipzig

SPIN 11812852   Printed on acid-free paper – 88/3100 – 5 4 3 2 1 0

Preface

The symposium "Artificial Economics 2006" is the second in a planned line of symposia on artificial economics, following a symposium held in Lille, France in 2005, organized by Phillipe Mathieu, Bruno Beaufils and Olivier Brandouy [1]. The organizing theme of these symposia, is the computational study of economies perceived as complex dynamic systems.

With the latter being a non-existing phenomenon, the defining distinction is not between *artificial* and *natural* economics, but rather between aiming to understand economic processes by constructively simulating them, as opposed to reductionistically analyzing economic systems. With this distinction the game is set, and doors are open for new understandings of economic systems.

Artificial economics is a methodological approach rather than a paradigmatic approach. Neoclassicals, Keynesians, Marxists etc. may all benefit from the methods of artificial economics. Surely some New Classicals have felt the straight jacket of eg. having to assume homogeneous or representative agents, and certainly many Keynesians have dreamt of unifying microeconomics and macroeconomics without totally giving up on their macromodel. Artificial economics provide a toolbox fit for turning towards such fundamental problems anew, without adopting a predetermined idea of what the answers are going to be.

What artificial economics does embrace is an encouragement to economics and economic subdisciplines, to take off the blinkers, and learn about other disciplines. Artificial economics encompasses implementation of ideas and modeltypes from other sciences into economics, integration of different economic submodels, as well as the export of economic conceptions to other sciences. The three invited speakers of Artificial Economics 2006, Akira Nametame, Thomas Lux and Kumaraswamy "Vela" Velupillai, together with a number of contributors, all prove that much may be gained by moving between disciplines.

Akira Nametame, from the Department of Computer Science, National Defense Academy, Yokosuka, Japan, has moved between the fields of physics, computer science and economics - or more generally, social sciences. With applied physics and operations research as his original fields, Nametame has in recent years commuted between economics and computer science, managing to enrich both fields with his interdisciplinary insights. In his speak, printed as Chapter 11 in this volume, Nametame will discuss the formation of social norms by means of interaction (network effects).

Thomas Lux, Department of Economics, University of Kiel, Germany, started his career in macroeconomics, but has made important contributions to finance by introducing new tools adapted from other sciences to the field. Among other contributions, he was one of the first to apply statistical mechanics to financial markets [3]. Following up on this theme, Lux has combined behavioural finance, agent-based computational economics and econophysics in order to explain the stylized facts of financial returns (eg. fat tails and volatility). In his speak Thomas Lux will discuss estimation of agent-based models.

Kumaraswamy "Vela" Velupillai, National University of Ireland, Galway, Ireland and Trento University, Italy, moves elegantly between several economic subdisciplines with macroeconomics as his home base, and a well-founded knowledge of mathematics, computability theory, philosophy etc. He is the founder of "Computable Economics" [2], i.e. a discipline in which results and theoretical tools stemming from classical recursion theory are applied to study fundamental economic problems with special reference to the computability, constructivity and complexity of economic decisions, institutions and environments. K. Velupillai has proven himself as a strong methodological watch dog - watching over both the analytical and the artificial approaches to economics, and this is also the position he shall take in his speak.

The Artificial Economics conferences are two-day symposia - a form that served its purpose well in Lille 2005 by generating interesting discussions between subfields - discussions that would not have arisen, had each subfield gone to different parallel sessions. The drawback is the limited number of papers that this form leaves room for. Again this year, space only permitted half of the submitted extended abstracts to be accepted. The difficult selection process was based on a double-blind reviewing process, where each paper was send to three referees. A thanks to all submitters of extended abstracts - without you there could be no symposium.

The Scientific Committee of Artificial Economics 2006 did a great job in reviewing submitted papers and broadcasting news about the Symposium. Thank You!

Scientic Committee:

- Frédéric AMBLARD - Université de Toulouse 1, France
- Robert AXTELL - Brookings Institute, USA
- Gérard BALLOT - ERMES, Université de Paris 2, France
- Bruno BEAUFILS - LIFL, USTL, France
- Giulio BOTTAZZI - S.Anna School, Pisa, Italy
- Paul BOURGINE - CREA, École Polytechnique, France
- Olivier BRANDOUY - CLAREE, USTL, France
- Charlotte BRUUN, Aalborg University, Denmark
- José Maria CASTRO CALDAS - ISCTE, DINAMIA, Portugal
- Christophe DEISSENBERG - GREQAM, France
- Jean-Paul DELAHAYE - LIFL, USTL, France
- Wander JAGER - University of Groningen, The Netherlands
- Marco JANSSEN - Arizona State University, USA
- Philippe LAMARRE - LINA, Université de Nantes, France
- Luigi MARENGO St. Anna school of advanced studies, Pisa, Italy
- Philippe MATHIEU - LIFL, USTL, France
- Denis PHAN - Université de Rennes I, France
- Juliette ROUCHIER - GREQAM, France
- Sorin SOLOMON - The Hebrew University of Jerusalem, Israel
- Leigh TESFATSION - Iowa State University, USA
- Elpida TZAFESTAS - National Technical University of Athens, Greece
- Murat YIDILZOGLU - IFREDE-E3i, Université Montesquieu Bordeaux IV, France
- Stefano ZAMBELLI - Aalborg University, Denmark

With special thanks to Bruno Beaufils, Olivier Brandouy and Philippe Mathieu, for putting their "symposium template" in my hands and guiding me all the way through, and to Stefano Zambelli for that little extra help and encouragement that means so much.

Aalborg,
June 2006

Charlotte Bruun

References

[1] Amblard, F., O. Brandouy and P. Mathieu (eds)(2005), Artificial Economics - Agent-Based Methods in Finance, Game Theory and Their Application. Lecture Notes in Economics and Mathematical Systems 564. Springer Berlin Heidelberg New York
[2] Velupillai, K. (2000) Computable Economics. Oxford University Press
[3] Lux, T. (1995) Herd Behaviour, Bubbles and Crashes. Economic Journal 105 1995 pp. 881-896

Contents

List of Contributors

Mikhail Anufriev
University of Amsterdam
The Netherlands

Alexandre T. Baraviera
UFRGS
Brasil

Ana L. C. Bazzan
UFRGS
Brasil

Andrew Bertie
Open University
UK

Olivier Brandoy
LEM, UMR CNRS-USTL
France

Andrea Consiglio
Università di Palermo
Italy

Silvio R. Dahmen
UFRGS
Brasil

S.A. Delre
University of Groningen
The Netherlands

Pietro Dindo
University of Amsterdam
The Netherlands

J.H. von Eije
University of Groningen
The Netherlands

Giorgio Fagiolo
University of Verona
Sant'Anna School of Advanced
Studies
Italy

César García-Díaz
University of Groningen
The Netherlands

Alexander Gorobets
Sevastopol National Technical
University
Ukraine

Cesáreo Hernández
University of Valladolid
Spain

Susan Himmelweit
Open University
UK

A.O.I. Hoffmann
University of Groningen
The Netherlands

Wander Jager
University of Groningen
The Netherlands

Toshiji Kawagoe
Future University
Japan

Valerio Lacagnina
Università di Palermo
Italy

Jacques Laye
LEF Inra Sae2/Engref
France

Maximilien Laye
Laboratoire d'Économétrie de l'École
Polytechnique
France

Marco LiCalzi
University of Venice
Italy

Charis Lina
University of Crete
Greece

Adolfo López-Paredes
University of Valladolid
Spain

Craig Lynch
Macquarie University
Australia

Philippe Mathieu
LIFL, UMR CNRS-USTL
France

Peter McBurney
University of Liverpool
UK

Alessio Moneta
Max Planck Institute of Economics
Germany

Thierry Moyaux
University of Liverpool
UK

Akira Namatame
National Defense Academy
Japan

Bart Nooteboom
Tilburg University
The Netherlands

J. Pajares
University of Valladolid
Spain

Stéphane Pajot
University of Rennes 1
France

Valentyn Panchenko
University of Amsterdam
The Netherlands

José A. Pascual
University of Valladolid
Spain

Paolo Pellizzari
U. of Venice
Italy

Denis Phan
University of Rennes 1
France

Marta Posada
University of Valladolid
Spain

Srinivas Raghavendra
National University of Ireland,
Galway
Ireland

Annalisa Russino
Università di Palermo
Italy

Shinichi Sasaki
Future University
Japan

Roberto da Silva
UFRGS
Brasil

Sitabhra Sinha
CIT Campus
India

Hervé Tanguy
Laboratoire d'Économétrie de l'École
Polytechnique
France

Andrew Trigg
Open University
UK

Paul Windrum
Manchester Metropolitan University
Business School
University of Maastricht
UK / The Netherlands

Arjen van Witteloostuijn
University of Durham
UK

Part I

Market Structure and Economic Behaviour

Heterogeneous Beliefs Under Different Market Architectures

Mikhail Anufriev[1] and Valentyn Panchenko[2]

[1] CeNDEF, University of Amsterdam, Amsterdam m.anufriev@uva.nl
[2] CeNDEF, University of Amsterdam, Amsterdam v.panchenko@uva.nl

Summary. The paper analyzes the dynamics in a model with heterogeneous agents trading in simple markets under different trading protocols. Starting with the analytically tractable model of [4], we build a simulation platform with the aim to investigate the impact of the trading rules on the agents' ecology and aggregate time series properties. The key *behavioral* feature of the model is the presence of a finite set of simple beliefs which agents choose each time step according to a fitness measure. The price is determined endogenously and our focus is on the role of the *structural* assumption about the market architecture. Analyzing dynamics under such different trading protocols as the Walrasian auction, the batch auction and the 'order-book' mechanism, we find that the resulting time series are similar to those originating from the noisy version of the model [4]. We distinguish the randomness caused by a finite number of agents and the randomness induced by an order-based mechanisms and analyze their impact on the model dynamics.

1.1 Introduction

The paper contributes to the analysis of the interplay between behavioral ecologies of markets with heterogeneous traders and institutional market settings. The investigation is motivated by the aim to explain inside a relatively simple and comprehensible model those numerous "stylized facts" that are left unexplained in the limits of the classical financial market paradigm (see e.g. [3]). Since the dynamics of financial market is an outcome of a complicated interrelation between behavioral patterns and underlying structure, it seems reasonable to start with an analytically tractable model based on realistic behavioral assumptions and to simulate it in a more realistic market setting. Such a strategy is chosen in this paper.

The first generation of agent-based models of financial markets followed the so-called bottom-up approach. The models were populated by an "ocean" of boundedly rational traders with adaptive behavior and were designed to be simulated on the computers. The Santa Fe artificial market (AM) model [1, 9] represents one of the best known examples of such approach. See also

[10] and reviews in [7] and [8]. The inherent difficulty to interpret the results of simulations in a systematic way led many researchers to build the models with heterogeneous agents which can be rigorously analyzed by the tools of the theory of dynamical systems. The achievements of the latter approach are summarized in [6]. In particular, the evolutionary model of Brock and Hommes (henceforth BH model) introduced in [4] follows the ideas of the Santa Fe AM in that the traders repeatedly choose among a finite number of predictors of the future price according to their past performance.

All the models mentioned so far (both simulational and analytic) are based on a simple framework with the mythical Walrasian auctioneer clearing the market. Real markets are functioning in a completely different way, and many recent models try to capture this fact. For instance, in [11] it is shown that an artificial market with a realistic architecture, namely an order-driven market under electronic book protocol, is capable of generating satisfactory statistical properties of price series (e.g. leptokurtosis of the returns distribution) in the presence of homogeneous agents. Similarly, the agent-based simulations in [2] demonstrate that the architecture bears a central influence on the statistical properties of returns. The latter contribution is also focused on the interrelation between market architecture and behavioral ecology, and in this respect is closely related to our paper. We relax, however, the assumption of a "frozen" population made in [2], and allow the agents to update their behavior over time.

More specifically, we assume that before the trading round, each agent can choose one of two simple predictors for the next price. The individual demand function depends on the predictor chosen, while the price is fixed later according to the specific market mechanism. The choice of predictor is implemented as a random draw with binary choice probabilities depending on the relative past performances of two predictors. An important parameter of the model is the intensity of choice, which measures the sensitivity of the choice probability to the relative performance. The higher the intensity of choice, the higher the probability that the best performing predictor is chosen. We simulate and compare the market populated by such heterogeneous agents under three aggregating mechanisms: Walrasian auction, batch auction, and an "order-book" mechanism. The latter two cases are interesting, since they resemble two protocols implemented in real stock exchanges. On the other hand, simulation of the Walrasian scenario provides a well-understood benchmark. Indeed, when the number of agents tends to infinity, our stochastic model converges to the deterministic BH model, thoroughly analyzed in [4].

In this paper, we show that understanding the basic mechanisms of the BH model can be very helpful also when dealing with more realistic market architecture. Indeed, the qualitative aspects of the non-linear dynamics generated by the BH model turn out to be surprisingly robust with respect to the choice of the market mechanism. Nevertheless, there are some important effects which realistic mechanisms supplement to the model. First, the finiteness of the number of agents provides a stabilizing effect on the model, since it implies a bigger noise in the choice of the predictor, which is equivalent to a smaller intensity of choice. Second, the inherent randomness of the markets under order-driven protocols (when agents have to choose one or few points from their demand curves) add destabilizing noise, which can be amplified, when the fundamental equilibrium is unstable. As a result, the generated time series remind the noisy version of the BH dynamics, when the system is switching between different attractors. This result is now produced, however, without adding either exogenous (e.g. due to the dividend realizations), or dynamic noise to the model. Third, we investigate the impact of two types of orders, market and limit orders, on the dynamics. We introduce a new parameter, the agents' propensity to submit market orders, which determines agent's preferences in submitting market orders as opposite to limit orders. We show that when this propensity high, the dynamics under the batch auction greatly deviate from the underlying fundamental, while the dynamics of the order-driven market converges to the dynamics under the Walrasian scenario. We also show some descriptive statistics for return time series generated for different values of the intensity of choice and the propensity to submit market orders.

The rest of the paper is organized as follows. In the next section we present the deterministic BH model, focusing on the agents' behavior, which is modeled in a similar way in our simulations. We also briefly discuss the properties of the dynamics for different values of the intensity of choice. In Section 1.3, we explain the three market mechanisms and introduce the difference between market and limit orders. Simulations results are presented and discussed in Section 16.5. Section 19.5 points to possible directions for future research.

1.2 The Brock-Hommes Benchmark Model

Let us consider a market where two assets are traded in discrete time. The riskless asset is perfectly elastically supplied at gross return $R = 1 + r_f$. At the beginning of each trading period t, the risky asset pays a random dividend y_t which is an independent identically distributed (i.i.d.) variable with mean $\bar{y}$. The price at period t is determined through a market-clearing condition (Walrasian scenario) and denoted by p_t. In the case of zero total supply of the risky asset, the fundamental price, which we denote by p_f, is given by the discounted sum of the expected future dividends $\bar{y}/r_f$. This is also the solution to the market-clearing equation for the case of homogeneous rational expectations.

In modeling the agents' behavior we closely follow the BH approach taken in [4]. Traders are mean-variance optimizers with absolute risk aversion a. Their demand for the risky asset reads

$$D_{i,t}(p_t) = \frac{E_{i,t-1}[p_{t+1} + y_{t+1}] - (1 + r_f)\,p_t}{a\,V_{i,t-1}[p_{t+1} + y_{t+1}]}, \qquad (1.1)$$

where $E_{i,t-1}[p_{t+1} + y_{t+1}]$ and $V_{i,t-1}[p_{t+1} + y_{t+1}]$ denote the expectations of trader i about, respectively, the mean and variance of price cum dividend at time $t + 1$ conditional upon the information available at the end of time $t - 1$. It is assumed that all the agents expect the same conditional variance σ^2 at any moment t, and that there are different predictors for the mean. Thus, the agents in the model have heterogeneous expectations.

We concentrate here on one of a few cases analyzed in [4] and assume that two predictors are available in the market, *fundamental* and *trend-chasing*. These two predictors capture, in a very stylized way, two different attitudes observed in real markets. The *fundamental* predictor forecasts the fundamental value $p^f = \bar{y}/r_f$ for the next period price, so that

$$E_t^1[p_{t+1} + y_{t+1}] = p^f + \bar{y}\,.$$

According to the trend-chasing predictor, the deviations from the fundamental price p^f can be persistent, i.e.

$$E_t^2[p_{t+1} + y_{t+1}] = (1 - g)\,p^f + g\,p_{t-1} + \bar{y}\,,$$

for some positive g.

In the BH model the population of agents is continually evolving. Namely, at the beginning of time t, agents choose one predictor among the two, according to their relative success, which in turn depends on the *performance measure* of predictors. The fraction n_t^h of the agents who use predictor $h \in \{1, 2\}$ is determined on the basis of the average profit π_{t-1}^h obtained by the traders of type h between periods $t - 2$ and $t - 1$. Since under the Walrasian market-clearing, all agents with a given predictor have the same profit, the average profit of a type in the BH model can be simply referred as the profit of a given type.

As soon as the profit π_{t-1}^h is determined, the performance measure U_{t-1}^h of strategy h can be computed. Agents have to pay a positive cost C per time unit to get an access to the fundamental strategy, and $U_{t-1}^1 = \pi_{t-1}^1 - C$, while the trend-chasing strategy is available for free, and hence, $U_{t-1}^2 = \pi_{t-1}^2$. In our simulation model, we, in addition, apply a transformation to this performance measure to make it scale-free: $\tilde{U}_{t-1}^h = U_{t-1}^h/(|U_{t-1}^1| + |U_{t-1}^2|)$. Finally, the fraction n_t^h is given by the discrete choice model, so that

$$n_t^h = \exp[\beta \tilde{U}_{t-1}^h]/Z_{t-1}\,, \quad \text{where} \quad Z_{t-1} = \sum_h \exp[\beta \tilde{U}_{t-1}^h]\,. \tag{1.2}$$

The key parameter β measures the *intensity of choice*, i.e. how accurately agents switch between different prediction types. If the intensity of choice is infinite, the traders always switch to the hystorically most successful strategy. On the opposite extreme, $\beta = 0$, agents are equally distributed between different types independent of the past performance.

Let us briefly discuss the dependence of the price dynamics on the intensity of choice in the BH model. For details the reader is refereed to [4], where

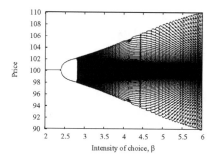 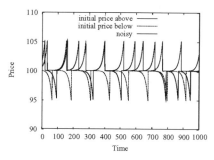

Fig. 1.1. Time-series properties of the Brock-Hommes model. **Left Panel:** Bifurcation diagram with respect to intensity of choice β. For each $\beta \in (2, 6)$, 500 points after 1000 transitory periods are shown for two different initial conditions: one below fundamental price and one above. The parameters are $C = 1$, $g = 1.2$, $r_f = 0.1$ and $\bar{y} = 10$. **Right Panel:** Typical time series for intensity of choice after the secondary bifurcation, in this case for $\beta = 4$. See text for explanation.

the deterministic skeleton with constant dividend is analyzed. From the bifurcation diagram shown in the left panel of Fig. 1.1, it can be seen that the fundamental equilibrium, where the price is equal to p_f, is stable for small values of β. For $\beta = \beta^* \approx 2.35$, a primary pitchfork bifurcation occurs, where the fundamental equilibrium loses stability. Two additional stable equilibria appear, one above and one below the fundamental and the original equilibrium becomes unstable. (Notice that for each β we show the prices for two initial conditions, belonging to the basins of attraction of two different equilibria.) A secondary Neimark-Sacker bifurcation takes place for $\beta = \beta^{**} \approx 2.78$. A stable quasiperiodic cycle emerges immediately afterward. With higher β the amplitude of this cycle increases, so that it almost touches the unstable fundamental equilibrium. For $\beta = \infty$ the system is close to a homoclinic bifurcation, which explains the typical time series for high β, reproduced in the right panel of Fig. 1.1.

If the initial price $p_0 > p_f$, then the price will grow (shown by solid thin line), further diverging from the unstable fundamental equilibrium. The trend following behavior, which is dominating due to its zero costs, is responsible for this market bubble. The forecasted error of trend-followers increases over time, however, since the actual price grows faster than expected. When the error becomes too high, it offsets the positive cost C of fundamental predictor. From this moment agents prefer to switch to fundamental behavior, contributing to a crash. From (1.2) it can be seen that, due to finite β, some small fraction of chartists remains in the market. This fact keeps the price a bit above the fundamental value and new bubble starts. A similar pattern with negative bubbles can be observed for initial price $p_0 < p_f$ (shown by the thin dotted line in the right panel of Fig. 1.1). Finally, if a small amount of dynamical noise is added, the positive and negative bubbles coexist on the trajectory (shown by the thick solid line). The observed behavior is qualitatively the same for all relatively high β, only the amplitude of the quasi-periodic cycle increases with β, as can be seen from the bifurcation diagram.

It is important to stress that the time series described above are obtained under the assumption of a constant dividend. Thus, the BH model is able to explain the excess volatility as an endogenous outcome of the agents' interactions. A more sophisticated model built in a similar spirit in [5] concentrates on the explanation of other stylized facts. The authors reproduce volatility clustering and realistic autocorrelation, kurtosis and skewness of the return distribution. Since the main goal of this paper is an investigation of the impacts of the market mechanisms on the model, but not the reproduction of the stylized facts, we will limit our analysis in the next sections to the simplest possible BH model.

1.3 Different Market Designs

On the basis of the analytic BH model, we construct an agent-based model and investigate its behavior under different trading protocols. In the agent-based model, the fraction n_t^h is interpreted as a probability of agent i to be of type h. The Walrasian auction is set as a benchmark, since the standard argument of the Law of Large Numbers implies that its outcome is equivalent to the original BH model as the number of agents tends to infinity. We will compare this setting with two more realistic order-driven markets, i.e. the batch auction and order book. Thought the paper we consider continuous prices.

1.3.1 Walrasian Auction

Under the Walrasian auction, at time t each agent i submits his excess demand function $\Delta D_{i,t}(p)$, which is the difference between his demand $D_{i,t}$ defined in (1.1) and his current position in the risky asset. The price p_t is determined from the market clearing condition $\sum_i \Delta D_{i,t}(p_t) = 0$. Notice that the equilibrium price p_t is always unique for the considered demand functions.

1.3.2 Batch Auction

Under the batch auction mechanism, each agent submits one or more orders, instead of the whole demand function. There are two types of the orders: limit and market order. A *limit order* consists of a price/quantity combination (p, q). Similarly to [2], an agent determines the price of a limit order as $p = p^* \pm \varepsilon |p_{t-1} - p^*|$, where p^* is the solution to the agent's "no-rebalancing condition" $\Delta D_{i,t}(p^*) = 0$, ε is a random variable, uniformly distributed on $[0, 1]$, and "+" corresponds to sell order and "−" to buy order. The quantity of the limit order at price p is given by $q = \Delta D_{i,t}(p)$. A *market order* specifies only the desired quantity of shares. As in [2], the type of order is determined by a propensity to submit a market order $m \in [0, 1]$, which is exogenously given parameter. A limit order (p, q) becomes a market order $(\cdot, q)$, if $\varepsilon < m$ in the limit order price equation. The price p_t is determined as an intersection of demand and supply schedules build on the basis of submitted orders (see [2] for details). Market buy/sell orders are priced at the min/max price among the corresponding side limit orders, which guaranties their fulfillment.

1.3.3 Order Book

In the order-book market, a period of time does not correspond to a single trade any longer. Instead, there is one trading session over period t and price p_t is the closing price of the session. Each agent can place only one buy or sell order during the session. The sequence in which agents place their orders is determined randomly.

During the session the market operates according to the following mechanism. There is an electronic book containing unsatisfied agents' buy and sell orders placed during current trading session. When a new buy or sell order arrives to the market, it is checked against the counter-side of the book. The order is partially or completely executed if it finds a *match*, i.e. a counter-side order at requested or better price, starting from the best available price. An unsatisfied order or its part is placed in the book. At the end of the session all unsatisfied orders are removed from the book.

As in the batch auction setting, there are two types of the orders: limit and market orders. The mechanisms for determining type of the order, its price and quantity are equivalent to those described in Section 1.3.2. The quantity of the market order is determined from the excess demand on the basis of the last transaction price.

1.4 Simulation Results

In Fig. 1.2, 1.3 and 1.4 we present the outcomes of typical simulations for different market architectures, different values of intensity of choice parameter β and different propensity to market orders. Ignoring transitory 1000 points, we show in each panel 4 time series, corresponding to the equilibrium price in the deterministic BH model (solid thick line), the equilibrium price in the simulated agent-based model under Walrasian (solid thin line) and batch (dashed line) auctions, and, finally, the closed price under order-book protocol (dotted line). Apart from the first two simulations, all the results are reported for 500 agents present in the market. For each β we compare the case $m = 0.1$, when nearly all orders are limit orders, with $m = 0.8$, when the majority of the orders are of market type. Finally, we consider five following values of intensity of choice. First, $\beta = 2.5$, which lies between two bifurcation values β^* and

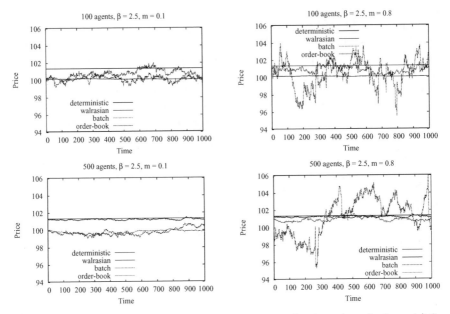

Fig. 1.2. Price time series under different market mechanisms (see the legends) for different number of agents and different propensity to submit market orders m (see the titles). Intensity of choice $\beta = 2.5$.

β^{**} (see Fig. 1.2). Then, $\beta = 2.75$ and $\beta = 2.8$, i.e. immediately before and after the secondary bifurcation (see Fig. 1.3). And finally, $\beta = 3$ and $\beta = 5$, i.e. far above β^{**}, when the quasi-periodic dynamics discussed at the end of Section 19.2 has already emerged (see Fig. 1.4).

For $\beta = 2.5$ the fundamental equilibrium is unstable, and the stable equilibrium of the BH model lies above $p_f = 100$, at the level $p^* \approx 101.3$. When the number of agents is small (as in the upper panels of Fig. 1.2), the discrepancy between the theoretical fraction of fundamentalists, n_t^f, computed according to (1.2) and the realized fraction is relatively large. Such discrepancy can be thought of as the agents' mistake in the computation of the performance measure. Therefore, it corresponds to a smaller "effective" intensity of choice with respect to $\beta = 2.5$. It explains why the relatively stable time series of Walrasian scenario lies well below the BH benchmark, close to $p_f = 100$: this is simply stable steady-state for some smaller value of β. When the number of agents increases, the error between the theoretical and realized fraction of fundamentalists decreases and the Walrasian scenario is getting closer to the BH benchmark (see the lower panels of Fig. 1.2).

The higher level of noise, which is intrinsic to the order-driven markets, has similar stabilizing consequences for the remaining two market mechanisms. This can be clearly seen in the lower left panel of Fig. 1.2, where price for both

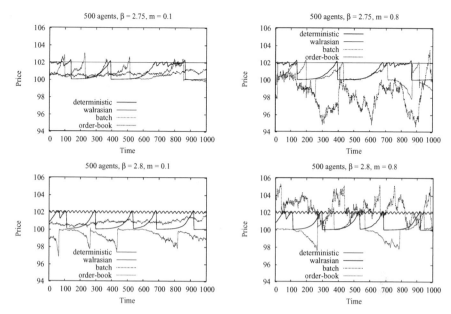

Fig. 1.3. Price time series under different market mechanisms (see the legends) for 500 agents, different intensities of choice and different propensity to submit market orders m (see the titles).

batch and order-book markets fluctuates around equilibrium, which is stable only for some smaller value of β. This stabilizing "β-effect" takes place also for other parametrizations, but usually cannot be seen, since other destabilizing effects dominate.

For example, in the right panels of Fig. 1.2, one can clearly see that an increase of the propensity to submit market orders m has strong destabilizing effect on the batch auction. It is interesting that the same increase of m has rather stabilizing consequences for the order-book mechanism and shifts the price towards the benchmark fundamental value (cf. 1.2, the two lower panels). This should not, however, come as a surprise, given the difference between these two mechanisms. Indeed, under the order-book, the executed prices of the market orders always come from some limit orders. Thus, the realized prices are still mainly determined by the limit orders, while increasing the randomness from the higher propensity to submit market orders m, probably leads to the stabilizing "β-effect" which we discussed above. On the other hand, under the batch protocol with many market orders, the price becomes very dependent on the relative sizes of buy and sell market orders and, therefore, its realization becomes more random by itself.

The two upper panels of Fig. 1.3 reveal another effect, implied by two types of randomness, i.e. one due to the errors between the theoretical and

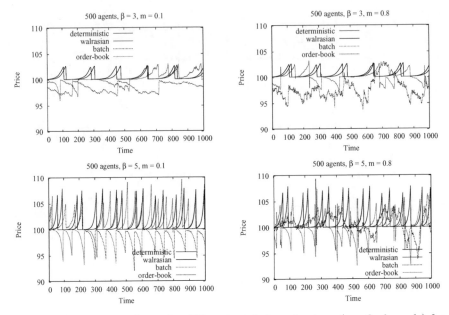

Fig. 1.4. Price time series under different market mechanisms (see the legends) for 500 agents, different intensities of choice and different propensity to submit market orders m (see the titles).

the realized fraction of traders, and one inherent in order-driven markets. Here, the BH model still generates stable dynamics converging to $p^* \approx 102$. The dynamics under the Walrasian auction and the order-book are unstable, however. The reason for this is a very small size of the basin of attractor p^*. The small endogenous noise constantly drives the dynamics out of this attractor, even if it ultimately comes back due to the instability of the fundamental fixed point. In addition, we again observe that "β-effect" has strong stabilizing effect for the batch auction with small propensity to submit market orders, $m = 0.1$. If the propensity is high, $m = 0.8$, the batch auction again leads to a very unstable behavior with large fluctuations and, sometimes, outliers. Similar characteristic can be given to the case $\beta = 2.8$, which is shown on the lower panels of Fig. 1.3.

Finally, Fig. 1.4 gives examples for relatively high values of β, when the stabilizing "β-effect" does not play a role, since the secondary bifurcation has already occurred under all market mechanisms. The main inference is that the analytical BH model based on the Walrasian auction is able to replicate the dynamics under more sophisticated trading mechanisms quite well. In particular, the time series in the two lower panels resemble the one obtained in the right panel of Fig. 1.1, when the dynamical noise triggers the dynamics between the two coexisting quasi-periodic attractors.

Table 1.1. Deceptive statistics of the return series generated under various market settings.

Auction	Walrasian	Batch $m = 0.1$	Order-Book	Walrasian	Batch $m = 0.8$	Order-Book
			$\beta = 2.50$			
mean	0.0000	0.0000	0.0000	0.0000	0.0000	0.0000
variance	0.0002	0.0005	0.0003	0.0002	0.0033	0.0004
skewness	−0.178	−0.040	−0.468	−0.178	−7.760	−0.033
kurtosis	0.357	0.046	1.153	0.357	123.384	0.631
			$\beta = 2.75$			
mean	0.0000	0.0000	0.0000	0.0000	0.0000	0.0000
variance	0.0007	0.0005	0.0012	0.0007	0.0027	0.0024
skewness	−12.693	−0.063	−0.410	−12.693	0.941	2.241
kurtosis	191.460	0.040	97.315	191.460	19.251	181.319
			$\beta = 2.80$			
mean	0.0000	0.0000	0.0000	0.0000	0.0000	0.0000
variance	0.0009	0.0005	0.0017	0.0009	0.0026	0.0019
skewness	−12.766	0.129	−9.356	−12.766	−2.141	13.710
kurtosis	185.355	0.073	118.671	185.355	22.256	407.478
			$\beta = 3.00$			
mean	0.0000	0.0000	0.0000	0.0000	0.0005	0.0000
variance	0.0014	0.0010	0.0021	0.0014	0.0269	0.0034
skewness	−13.151	−12.891	−10.871	−13.151	−10.757	−0.471
kurtosis	183.716	234.023	138.080	183.716	243.943	103.755
			$\beta = 5.00$			
mean	0.0000	0.0000	0.0001	0.0000	0.0000	0.0001
variance	0.0059	0.0054	0.0091	0.0059	0.0030	0.0133
skewness	−10.602	−7.965	−4.988	−10.602	0.199	1.265
kurtosis	115.248	66.308	47.613	115.248	13.820	43.380

Table 1.1 shows descriptive statistics of the return series for various β and m under different market auctions. In most cases the values of the skewness and kurtosis are far from realistic (e.g. S&P series returns statistics reported in [5]). Nevertheless, for $\beta = 5$ and $m = 0.8$ the values of the statistics for the batch and order-book auctions become closer to the realistic values.

1.5 Conclusion

The analytically tractable BH model introduced in [4] is quite successful in reproducing a number of stylized facts. Indeed, when the intensity of choice in this model is high, the price time series may deviate from fundamental benchmark in a systematic way, become quasi-periodic or even chaotic, and exhibit excess volatility. The phenomenon of volatility clustering can also be reproduced in a similar framework, as discussed e.g. in [5]. However, the unrealistic market clearing scenario, where each agent has to supplement an infinite amount of information to an (in)famous Walrasian auctioneer, has always cast a shadow on such an explanation of the stylized facts.

The results of this paper suggest that the order-based model is able to replicate the main features of the evolutionary BH model. Moreover, we found that the finiteness of the number of agents provides stabilizing effect, which is equivalent to a lower intensity of choice β in the deterministic model. The randomness resulting from the batch auction and the order-book mechanism destabilizes the model. This effect is mainly observed when the basins of attraction of the steady state (cycle) are small, i.e. in the vicinity of a bifurcation.

While investigating the effects of the limit- and market order, we found that the presence of the large number of market orders may substantially destabilize the dynamics of the batch auction. Instead, under the book-order mechanism, this effect is not observed.

The analysis of the descriptive statistics of the return series for different parameters and under different market protocols suggests that the structural assumptions are able to explain only some stylized facts, e.g. excess kurtosis. The model did not generate volatility clustering under any protocol, which suggest that this phenomenon should be modeled using the appropriate behavioral assumptions.

This result brings us to the directions for the future research. It would be interesting to start with a more realistic model (e.g. the model [5]), which is able to reproduce volatility clustering, and investigate its dynamics under various market mechanisms. Moreover, we could adopt different mechanisms for the limit order price generation, which are closer to those observed on the real markets. On the behavioral level, we could distinguished some parameters (e.g. β) between agents within one group and introduce a memory parameter into the individual type selection procedure.

References

[1] Arthur WB, Holland JH, LeBaron B, Palmer R, Tayler P (1997) Asset pricing under endogenous expectations in an artificial stock market. In: Arthur WB, Durlauf SN, Lane D (eds) The economy as an evolving complex system II: 15-44. Addison-Wesley

[2] Bottazzi G, Dosi G, Rebesco I (2005) Institutional architectures and behavioral ecologies in the dynamics of financial markets: a preliminary investigation. Journal of Mathematical Economics 41: 197-228

[3] Brock WA (1997) Asset price behavior in complex environments. In: Arthur WB, Durlauf SN, Lane DA (eds) The economy as an evolving complex system II: 385-423. Addison-Wesley

[4] Brock WA, Hommes CH (1998) Heterogeneous beliefs and routes to chaos in a simple asset pricing model. Journal of Economic Dynamics and Control 22: 1235-1274

[5] Gaunersdorfer A, Hommes CH (2005) A nonlinear structural model for volatility clustering. In: Teyssière G, Kirman A (eds) Long Memory in Economics: 265-288. Springer

[6] Hommes CH (2006) Heterogeneous agent models in economics and finance. In: Judd K, Tesfatsion L (eds) Handbook of Computational Economics II: Agent-Based Computational Economics. Elsevier, North-Holland

[7] LeBaron B (2000) Agent-based computational finance: suggested readings and early research. Journal of Economic Dynamics and Control 24: 679-702

[8] LeBaron B (2006) Agent-based computational finance. In: Judd K, Tesfatsion L (eds) Handbook of Computational Economics II: Agent-Based Computational Economics. Elsevier, North-Holland

[9] LeBaron B, Arthur WB, Palmer R (1999) Time series properties of an artificial stock market. Journal of Economic Dynamics and Control 23: 1487-1516

[10] Levy M, Levy H and Solomon S (2000) Microscopic simulation of financial markets. Academic Press, London

[11] LiCalzi M, Pellizzari P (2003) Fundamentalists clashing over the book: a study of order-driven stock markets. Quantitative Finance 3: 1-11

The Allocative Effectiveness of Market Protocols Under Intelligent Trading

Marco LiCalzi[1] and Paolo Pellizzari[2]

[1] Dept. Applied Mathematics and SSAV, U. of Venice, Italy `licalzi@unive.it`
[2] Dept. Applied Mathematics and SSAV, U. of Venice, Italy `paolop@unive.it`

2.1 Introduction

An important criterion for the evaluation of an exchange market is its ability to achieve allocative efficiency. The seminal paper by Gode and Sunder (1993) shows that the protocol known as continuous double auction can attain the efficient allocation even if the traders exhibit "zero-intelligence": hence, market protocols may actively contribute to the discovery of an efficient allocation. This paper spawned a variety of computer simulations that "enabled us to discover that allocative efficiency [...] is largely independent of variations in individual behavior" at least in canonical environments; see Sunder (2004).

However, the attainment of allocative efficiency is only a necessary condition for the effectiveness of a market protocol in an exchange economy. For instance, consider the fictitious protocol of Walrasian tâtonnement, where a centralized market maker iteratively elicit traders' excess demand functions and adjust prices before trade takes actually place. Under standard conditions, this protocol attains allocative efficiency while simultaneously minimizing both the volume of transactions and price dispersion. Moreover, the efficient allocation is reached in one giant step, so that its speed of convergence (after trade begins) is instantaneous.

Clearly, the Walrasian mechanism is only an idealization. Realistic market protocols require far less information from traders and should not be expected to perform as smoothly. This raises the question of ranking the effectiveness of those different market protocols which are commonly used in real markets; see Audet et al (2002) or Satterthwaite and Williams (2002). Assuming that they all pass the test of achieving an efficient allocation, which additional criteria should enter in their comparison? Walrasian tâtonnement suggests at least three possibilities: excess volume, time to convergence, and price dispersion.

A major complication in the study of alternative protocols is that their outcome is profoundly affected by traders' behavior; see Brewer et al (2002). This may exhibit sophisticated strategies, behavioral biases, access to different forecasting abilities, and a variety of factors which we encompass under the

term of traders' *intelligence*. Gode and Sunder (1993) introduced the notion of
"zero intelligence" as an extreme assumption, under which all complications
in traders' behavior are ruled out and traders are only requested to satisfy a
natural budget constraint. They argued that the outcome of a market protocol
under zero intelligence is a test of its intrinsic ability to perform effectively.

Assuming zero intelligence, LiCalzi and Pellizzari (2005) compares the performance of different market protocols with regard to allocative efficiency and
other criteria such as excess volume or price dispersion. The main protocols
examined are: the batch auction, the continuous double auction, a (nondiscretionary) specialist dealership, and a hybrid of these last two. All the four
protocols exhibit a remarkable ability to achieve allocative efficiency under
three variants of zero intelligence, confirming the main insight from Gode and
Sunder (1993).

However, even under zero intelligence, stark differences in performance
emerge over other relevant dimensions. The continuous double auction has
the worst performance with respect to excess volume, time to convergence,
and price dispersion. The dealership has a lower time to convergence and
never performs worse than the batch auction. These differences are sometimes
dramatic and sometimes small (but persistent). Hence, LiCalzi and Pelllizzari
(2005) concludes that (under zero intelligence) there is a clear partial ranking
of these protocols with respect to excess volume, time to convergence, and
price dispersion. A dealership performs slightly better than a batch auction or
a hybrid market, and both are substantially more effective than a continuous
double auction.

The relevance of this conclusion for the evaluation of practical market protocols is severely limited by the assumption of zero intelligence, which rules
out the impact of differences in traders' behavior. The question addressed
in this paper is how much of this conclusion remains true if we remove zero
intelligence. Using two simple rules for intelligent trading, we study the performance ranking for the four market protocols with regard to excess volume,
time to convergence, and price dispersion.

The organization of the paper is the following. Section 2.2 describes the
model used in our simulations. Section 2.3 details the experimental design.
Section 2.4 reports on the results obtained and Section 2.5 offers our conclusions. For an expanded and more robust analysis, see LiCalzi and Pellizzari
(2006).

2.2 The Model

We use the same setup as in LiCalzi and Pellizzari (2005), where a simple
exchange economy admits a unique efficient allocation. Given that the market protocols attain allocative efficiency, this implies convergence to the same
allocation and facilitates comparisons. Following Smith (1982), we identify

three distinct components for our (simulated) exchange markets. The environment in Section 2.2.1 describes the general characteristics of the economy, including agents' preferences and endowments. The market protocols in Section 2.2.2 provide the institutional details which regulate the functioning of an exchange. The behavioral assumptions in Section 2.2.3 specify how agents make decisions and take actions.

2.2.1 The Environment

We consider an economy with n traders. There is cash and one good, which is an asset with a (random) realization value Y at a given time T in the far future. Each trader i has an initial endowment of cash $c_i \geq 0$ and shares $s_i \geq 0$. We rule out any informational effect and assume that all traders believe that Y is normally distributed with mean $\mu \geq 0$ and precision $\tau = 1/\sigma^2 > 0$ and that no new information is ever released. Therefore, traders' beliefs about Y are homogeneous and never change until uncertainty resolves.

Each trader i has CARA preferences over his final wealth, with a coefficient of risk tolerance $k_i > 0$. Therefore, trader i's excess demand function for the asset (net of his endowment s_i) is the linear function

$$q_i(p) = k_i \tau (\mu - p) - s_i. \tag{2.1}$$

Let $K = \sum_i k_i$ be the sum of traders' coefficients of risk tolerance. The unique efficient risk-sharing allocation for this economy requires that trader i holds $s_i^* = (S/K)k_i$ asset shares; that is, it is proportional to the coefficient of risk tolerance. The competitive equilibrium achieves the efficient allocation at the price $p^* = \mu - S/(\tau K)$. At such price, the trader i's net demand $q_i(p^*)$ is exactly filled, making his final allocation $q_i(p^*) + s_i$ equal to the required $s_i^* = (S/K)k_i$.

2.2.2 The Market Protocols

We compare the performances of four market protocols: a batch auction, a continuous double auction, and a nondiscretionary dealership, and a hybrid of these last two. The first protocol is simultaneous, while the other three are sequential. The following features are common to all protocols.

A protocol is organized in trading sessions (or days). Agents participate in every trading session, but each of them can exchange at most one share per session. If the protocol is sequential, the order in which agents place their orders is randomly chosen for each trading session. If the protocol is simultaneous, all order are made known and processed simultaneously so the time of their submission is irrelevant. In every trading session, each agent selects on which side of the market he attempts to place a trade: he can switch roles across trading sessions, but he cannot place simultaneous orders for buying and selling within the same session. The books are completely cleared at the end of each trading session.

Prices are quoted using a minimum tick; in other words, they are discretized. Moreover, prices must be nonnegative: if a trader places a bid lower than zero, this is ignored; if a trader places an ask lower than zero, this is automatically converted to the lowest strictly positive price compatible with the existing tick.

The specific market protocols studied in this paper are the following.

Batch auction. In each trading session, after traders submit their orders, the exchange price p^* is obtained at the intersection of demand and supply. If there are multiple solutions, we choose p^* as the midpoint of the interval between the lowest and the highest possible values. (If there are no solutions, no exchange takes place.) Shares and corresponding payments are exchanged between traders who submitted bids not lower than p^* and asks not higher than p^*. Traders who placed orders exactly at price p^* may be accordingly rationed. This protocol is also known as the k-double auction, with $k = 1/2$.

Continuous double auction. In each trading session, traders place their orders on the selling and buying books. Their orders are immediately executed if they are marketable; otherwise, they are recorded on the books with the usual price-time priority. Orders are canceled only when a matching order arrives or the trading day is over.

Nondiscretionary dealership. There is a specialist dealer who posts bid and ask quotes valid only for a unit transaction. Agents check sequentially the dealer's quotes for the side of the transaction they are attempting. If an agent accepts the dealer's quote, the exchange takes place at the quoted price. Right after a transaction is completed, the two dealer's quotes for bid and ask increase (or decrease) by one tick if the agent completed a purchase (or a sale). The size of the bid-ask spread stays fixed over time, so the price is never unique. Limited to this protocol, therefore, convergence of prices to a given value p^* should be interpreted as convergence to within a bid-ask interval that contains p^*.

Hybrid market. This combines the continuous double auction with the dealership. Distinct selling and buying books hold quotes from the specialist dealer and from the public, respectively. The dealer posts bids and asks valid only for a unit transaction and revises her quotes as in the nondiscretionary dealership; in particular, she moves her quotes only after transactions in which she has been involved. Agents check sequentially the books for the side of the transaction they are attempting. Their orders are immediately executed at the best price available (which may be different from the specialist's) if they are marketable; otherwise, they are recorded on the traders' book with the usual price-time priority. Agents' orders are canceled only when a matching order arrives or the trading day is over. Hence, once deposited on the traders' book, an order from an agent cannot be executed with the dealer.

2.2.3 Behavioral Assumptions

A major obstacle in the study of microeconomic systems is that their performance is jointly determined by the interactions of traders' behavior within the market protocol. As traders may react differently to different market protocols, it is difficult to separate the intrinsic characteristics of a market protocol from the properties induced by the traders' strategies. Our approach is to concentrate on the institutional characteristics of the protocols, by making general-purpose assumptions on traders' behavior. These assumptions hold for all the simulations reported in this paper.

First, traders are restricted to trade one unit at a time. This restriction on traded quantities simplifies the strategy space and allows direct comparisons with existing theoretical results. Second, buying orders are constrained by the available cash and selling orders by the available endowment of the asset; that is, budget constraints hold. This is consistent with a value-based strategy ("buy low, sell high"), which is a seemingly natural requirement of rationality for traders' behaviors.

Third, since the demand function (2.1) of each trader is strictly decreasing, traders have decreasing marginal utility for additional units. If the current endowment of a trader is s_i, his valuation for the next unit to trade is

$$p(\pm 1) = \mu - \frac{s_i \pm 1}{k_i \tau},\tag{2.2}$$

where the $\pm$ sign depends on whether the attempted trade is a purchase or a sale. Clearly, this implies that the reservation price of each trader depends on the side of the transaction he is entering and on his current endowment s_i; moreover, his (implicit) bid-ask spread is $2/(k_i \tau)$.

Given his valuation, in each stage a trader must decide which side of the transaction he wants to attempt and what price to offer. These two separate decisions may exhibit various degrees of intelligence. LiCalzi and Pellizzari (2005) models zero intelligence as follows. At the start of a trading session, each trader chooses either side with equal probability. This randomized choice is stochastically independent of previous history, endowment, or any other parameter of the model. Hence, a trader ignores that the current market price is an imperfect signal for whether he should seek to buy or sell. After the trader has chosen his side of the transaction, suppose that he is going to attempt a purchase. Then his valuation for the next unit to buy is $p(+1)$ from Equation (2.2). Under zero intelligence, this (potential) buyer bids a price uniformly drawn from the interval $[p(+2), p(+1)]$. Similarly, a (potential) seller asks a price uniformly drawn from the interval $[p(-1), p(-2)]$. Again, the information associated with the current market price is ignored.

There are numerous possibilities to make traders "intelligent", ranging from the simple to the highly sophisticated. We attempt to capture the essence of intelligent trading by making two distinct assumptions that exploit the imperfect signal associated with the current price. One is concerned with the choice of which transaction to attempt, and the other with the price that is offered or asked.

Suppose that the current market price[3] is p. An intelligent trader decides the side of his next (potential) transaction by comparing p with his current valuations $p(+1)$ and $p(-1)$. If $p(+1) > p$, the market price is lower than the price at which the trader would like to buy one more unit, so he attempts a purchase. If $p(-1) < p$, the market price is higher than the price at which the trader would like to sell one more unit, so he attempts a sale. If $p(-1) \leq p \leq p(+1)$, the probability that he attempts a purchase is proportional to the distance of p from $p(+1)$. Formally, we assume that with probability

$$P(\text{sale}) = \begin{cases} 0 & \text{if } p(-1) < p \\ \frac{p-p(+1)}{p(-1)-p(+1)} & \text{if } p(-1) \leq p \leq p(+1) \\ 1 & \text{if } p(+1) > p \end{cases}$$

the trader attempts a sale and otherwise goes for a purchase. For later use, we nickname this assumption S_1 as a mnemonic for "side". The former zero intelligence assumption with $P(\text{sale}) = 1/2$ is denoted S_0. (The subscripts "1" and "0" denote the presence or absence of intelligence.)

Next, consider the choice of the price. Suppose that the trader is attempting a purchase. Under zero intelligence, he would post a bid uniformly drawn from the interval $[p(+2), p(+1)]$. We model intelligent trading by assuming that he compares the current market price p and his demand function to find the interval $[p(n + 1), p(n)]$ which contains p and then posts a bid uniformly drawn from $[p(n + 1), p(1)]$. Compared to zero intelligence, this trader selects his bid from a larger and more aggressive interval. There is a nice intuition for this rule: at a price p in $[p(n + 1), p(n)]$, the trader would like to buy n units. However, as we constrain him to buy one unit a time, he can at best try to buy the next unit at a price no greater than $p(n + 1)$. The symmetric version holds when the trader is attempting a sale. We nickname this rule P_1 as a mnemonic for "price"; P_0 denotes the rule under zero intelligence.

2.3 Experimental Design

2.3.1 Identification

A simulation run for our model requires the specification of five global parameters, a list of individual variables for each trader, as well as specific assumptions about market protocol and traders' behavior. The global parameters are the number n of traders, the mean μ and the variance σ^2 of the realization value Y of the asset, the number t of trading sessions, and the size Δ of

---

[3] For a batch auction, we use the price observed in the last active trading session. For the sequential protocols, we use the midpoint of the (best available) bid-ask spread.

the tick. Individually, a trader i is characterized by his coefficient k_i of risk tolerance and by his endowment of cash c_i and asset shares s_i. Finally, for protocols involving the dealer, we need to select her initial quotes.

The market protocols are described in Section 2.2.2. For ease of reference, we nickname the protocols as B (batch auction), C (continuous double auction), D (automated dealership), and H (hybrid market). Recall that the behavioral assumptions described in Section 2.2.3 are nicknamed S_i and P_k for $i, k = 0, 1$.

We have run simulations for all $4 \times 4 \times 3 = 48$ possible combinations of protocols, behavioral assumptions and performance criteria, over different instantiations of the parameters. The results reported in Section 2.4 are robust both to variations in the fine details in the market protocols and substantial changes in the parameters, provided that the overall liquidity of the system is sufficiently large. To simplify the presentation, we fix the exemplar parametric configuration reported in Table 2.1 and for each performance criterion we report the simulations for the four market protocols and the four behavioral assumptions. The initial dealer's quotes are a bid of 745 and an ask of 751, with a fixed bid-ask spread of 6. The competitive equilibrium price is $p^* = \mu - 2\sigma^2 = 760$ in all the simulations reported in this paper.

Table 2.1. Exemplar for identification.

	Parameters		Initialization
Global	n	$=$	$1,000$
	μ	$=$	$1,000$
	σ^2	$=$	120
	t	$=$	$2,500$
	Δ	$=$	1
Trader	k_i	$=$	divisors of σ^2 in $\{10, \ldots, 40\}$
	c_i	$=$	$50,000$
	s_i	$=$	permutation of $2k_i$

We say that a market protocol *exactly implements* a trading rule if it is never necessary to round traders' offers to match the ticked prices; see LiCalzi and Pellizzari (2005). An exact implementation allows *exact* convergence to the equilibrium price supporting the efficient allocation. (This is not relevant for the dealership protocol, because the fixed bid-ask spread prevents the price from being unique.) Our exemplar case is chosen to ensure that all the simulations reported in this paper satisfy the requirement of exact implementation. To this purpose, we choose integer values for μ and σ^2 and initialize each k_i's with a stochastically independent draw from a uniform distribution over the divisors of σ^2. We also assume that the support of the uniform distribution over bids and asks is formed by the integers in the two intervals.

2.3.2 The Simulations

A round of testing requires to simulate $3 \times 4 = 12$ combinations of performance criteria and behavioral assumptions. A typical round of simulations runs as follows. For each of the 12 combinations, we instantiate parameters according to the exemplar in Table 2.1 and work out a simulation batch consisting of 100 runs under different initial random seeds. Then we record the time series for prices, volume, and endowments, and compute relevant statistics for the performance criteria. The simulations have been run using a dedicated package of routines written in Pascal.

2.4 Results

We separately evaluate the performance of the four market protocols with respect to three criteria: excess volume, time to convergence, and price dispersion. Each one is defined and discussed in one of the following three subsections.

2.4.1 Excess Volume

Getting from the initial endowment to the efficient allocation requires a minimum number of (unit) transactions. The traded volume is the total number of unit transactions completed before attaining the efficient allocation. The Walrasian protocol attains the efficient allocation in one step and thus minimizes the traded volume. Realistic market protocols usually waste transactions and thus require higher volumes. We measure the *excess volume* in a market protocol as the percentage of traded volume in excess of the minimum required to attain the efficient allocation. Clearly, higher excess volumes signal less effective protocols that let unnecessary trades take place.[4]

Figure 2.1 shows four boxes. Each box is associated with a different assumption about the intelligence of the traders, as noted at its bottom. For instance, the top-right box is associated with $S_1 P_0$: this corresponds to positive intelligence in the choice of the side and zero intelligence in the price decision. Within each box, we graph the excess volumes for 100 runs for each of the four protocols, as well as marking the average level. The dots are color-coded: black is Batch (B), red is Continuous Double Auction (C), green is Dealership (D), and Blue is the Hybrid (H) protocol. The market protocols perform quite differently and these differences persist under various forms of trading intelligence.

Figure 2.2 merges the dots from the four boxes of Figure 2.1 in a single box. Two main findings emerge. First, the batch auction and the dealership

---

[4]   When a dealer mediates the transfer of one unit from a trader to another one, we record only one transaction so that the statistics for excess volume are directly comparable.

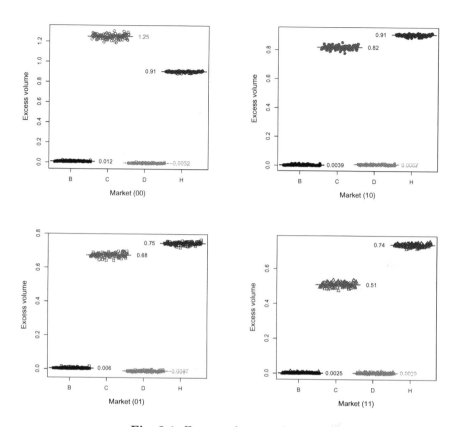

Fig. 2.1. Excess volume — datapoints.

have a substantially lower excess volume than the other two protocols under any of our variants of intelligent trading; regardless of these, the former two never exhibit more than 4% excess volume, while the latter two never go below 40%. Second, increasing trading intelligence tends to reduce excess volume, most notably in the continuous double auction, but does not eradicate the differences.

We fit a linear model to the data using a robust regression based on an M estimator; see Venables and Ripley (2002). The independent dummy variables are B, C, D, and H for the protocols, and P, S for trading intelligence over price and side. Dummies for protocols are increasingly ordered by the size of their effect: here, we leave out D because it has the lowest marginal impact. With t-values reported below each coefficient, the estimated equation for the excess volume is

$$\text{ExcVol} = \underset{(16.70)}{0.0566} + \underset{(1.673)}{0.0065B} + \underset{(183.9)}{0.7192C} + \underset{(209.8)}{0.8205H} - \underset{(-9.126)}{0.0252S} - \underset{(-32.13)}{0.0889P}$$

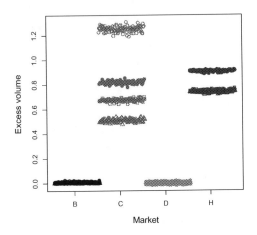

Fig. 2.2. Excess volume — merged datapoints.

The intercept, of course, combines the joint effect of D and $S_0 P_0$ and thus the baseline is a dealership protocol with zero intelligence trading. It is clear that the (average) effect of C and H on increasing the excess volume is statistically significant. Similarly, side and price intelligence decrease the excess volume.

2.4.2 Time to Convergence

Our second performance criterion is the number of trading sessions completed before no further trading takes place. In our exemplar case, the maximum number of units between the initial endowment and the final efficient endowment is 60, so this is a lower bound on the number of trading sessions required to achieve allocative efficiency. Figure 2.3 is similar to Figure 2.2 and reports the merged datapoints for time to convergence.

The estimated equation for time to convergence is

$$\text{Time} = \underset{(45.99)}{183.31} + \underset{(3.481)}{16.022H} + \underset{(34.06)}{156.78C} + \underset{(45.40)}{208.98B} - \underset{(-45.86)}{149.26S} + \underset{(2.311)}{7.5212P}$$

The (average) effect of H, C and B on increasing the time to convergence is statistically significant. Remarkably, while side intelligence contributes to this reduction, the coefficient for price intelligence denotes a (weak) contrary effect — when trading is aggressive, time to convergence lengthens.

2.4.3 Price Dispersion

Our third and final performance criterion attempts to quantify the dispersion of prices by measuring the standard deviation of the time series of the prices

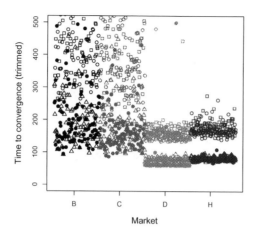

Fig. 2.3. Time to convergence — merged datapoints.

observed at the end of each trading session. Figure 2.4 reports the merged datapoints for price dispersion.

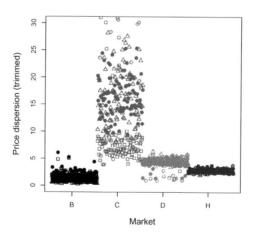

Fig. 2.4. Price dispersion — merged datapoints.

The estimated equation for price dispersion is

$$\text{PDisp} = \underset{(29.61)}{1.5128} + \underset{(23.50)}{1.3868H} + \underset{(52.32)}{3.0874D} + \underset{(261.0)}{15.398C} + \underset{(6.773)}{0.2826S} - \underset{(-12.05)}{0.5027P}$$

The (average) effect of H, D and C on increasing price dispersion is statistically significant, and particularly sizeable for C. Side intelligence has also an increasing effect, while price intelligence has a moderating impact.

2.5 Conclusions

The simulations shows that the choice of a protocol may have a substantial impact on the allocative effectiveness of an exchange market. Since lack of space prevents us from a longer analysis, we offer only the main conclusions. A richer and more complete analysis is carried out in LiCalzi and Pellizzari (2006).

Excess volume. The ranking with respect to excess volume is $\{B, D\} >>$ $C > H$, where > stands for "lower volume" and >> for "much lower volume". The notation $\{B, D\}$ means that the ranking is not statistically significant. In simple words, the batch auction and the dealership generate minimal excess volume; on the other hand, protocols involving a continuous double auction are seriously wasteful. Moreover, intelligent trading helps, in the sense that increasing the intelligence of traders tends to reduce (but does not eradicate) the excess volume.

Prescriptively, this suggests that a market regulator attempting to reduce excess volume in an exchange market would be well advised to opt for a batch auction or a dealership. Moreover, he should make an effort to educate traders towards making use of the signals embedded in the market price.

Time to convergence. The ranking with respect to time to convergence is $D > H >> C > B$, where > stands for "lower time" and >> for "much lower time". Protocols involving a dealer converge much faster. Intelligent trading is overall beneficial but has an ambiguous effect. A better choice for the side of the transaction to attempt substantially reduces the time to convergence: this alone might wipe out differences among all protocols except for the batch auction. On the other hand, more aggressive behavior on the choice of the prices slightly increases this time.

Prescriptively, this suggests that a market regulator attempting to reduce the time to convergence in an exchange market should consider having a dealership (possibly along an open book). Moreover, he should point out to traders the importance of using the price signal to understand the direction in which trade should be oriented, while attempting to reduce their greediness.

Price dispersion. The ranking with respect to price dispersion is $B > H >$ $D >> C$, where > stands for "lower dispersion" and >> for "much lower dispersion". The batch auction minimizes price dispersion and the continuous double auction yields by far the worst performance in this respect. Intelligent trading is overall damaging but with an ambiguous effect. More intelligence on choice of the side of the transaction increases the dispersion, while a more aggressive pricing behavior has a mild moderating effect. Prescriptively, this suggests that a market regulator attempting to reduce price dispersion in an exchange market should avoid the use of a continuous double auction.

References

[1] Audet N, Gravelle T, Yang J (2002) Alternative trading systems: Does one shoe fit all?. Working paper 2002-33, Bank of Canada, November

[2] Brewer PJ, Huang M, Nelson B, Plott CR (2002) On the behavioral foundations of the law of supply and demand: Human convergence and robot randomness. Experimental Economics 5:179-208

[3] Gode DK, Sunder S (1993) Allocative efficiency of markets with zero intelligence traders: Market as a partial substitute for individual rationality. Journal of Political Economy 101:119–137

[4] LiCalzi M, Pellizzari P (2005) Simple market protocols for efficient risk sharing. RePEc:wpa:wuwpfi:0504019, April.

[5] LiCalzi M, Pellizzari P (2006) The allocative effectiveness of simple market protocols. Working paper.

[6] Satterthwaite MA, Williams SR (2002) The optimality of a simple market mechanism. Econometrica 70:1841–1863

[7] V. L. Smith (1982), "Microeconomic systems as an experimental science", *American Economic Review* **72**, 923–955.

[8] Sunder S (2004) Markets as artifacts: Aggregate efficiency from zero intelligence traders. In: Augier ME, March JG (eds), Models of Man: Essays in Memory of Herbert A. Simon. The MIT Press, 501–520

[9] Venables WN, Ripley BD (2002) Modern Applied Statistics with S. Fourth edition. Springer.

Strategic Behaviour in Continuous Double Auction

Marta Posada, Cesáreo Hernández, and Adolfo López-Paredes

University of Valladolid, E.T.S. de Ingenieros Industriales, Paseo del Cauce s/n, 47011 Valladolid, Spain posada@eis.uva.es

Summary. We analyze with a bottom-up approach the competition between artificial intelligent agents in Continuous Double Auction markets in terms of allocative efficiency, price convergence and emergence or not of Nash equilibriums. In previous works agents have a fixed bidding strategy during the auction, usually under symmetric environments. In our simulations we allow the soft-agents to learn not only about how much they should bid or offer, but also about possible switching between the alternative strategies. We examine the behaviour of strategic traders under general supply and demand schedules (asymmetric environments) thus extending previous results.

The results clarify the limitations and the scope of Gode and Sunder conjecture and related recent works, and show the emergence of Hayeks and A. Smith endogenous order. Institutions matter and so does intelligence as far as the rate of convergence and agents surplus is concerned. These results are of importance in the design and performance of auctions in the real world and in the applications of auction theory to many problems in management and production, far beyond market design (market oriented programming).

3.1 Beyond Experimental Economics: a Research Agenda in Management Engineering (Scope and Related Work)

Following Vernon Smith [12] and other experimental economists, there are three dimensions that are essential in the design of any market experiment (figure 3.1): the institution (I) (it is both the exchange rules and the way the contracts are closed, and the information network), the environment (E) (agent endowments and values, resources, knowledge) and the agent behaviour (A).

By mapping different arrangements of the elements of this triplet (IxExA) into observed and forecasted outcomes, a host of experimental results can be obtained. For the last 30 years experimental economists have been doing just that and they have gained an accepted reputation when auction theory has to face practice. In the FCC auction design, for example, successful tests were

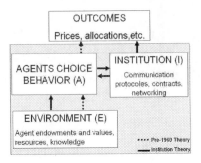

Fig. 3.1. Dimensions in the design of any market

conducted by Charles Plott in his laboratory at Caltech and help convince the FCC to adopt the theoretical proposals of Milgrom-Wilson design. Experimental Economics is now the companion of mechanistic design when defining auction institutions in practice.

But if we want to control the agent behaviour (A) dimension of our experiments, we have to move from human to artificial agents as argued in López and Hernández [8]. Taking this step, a reach program of research comes up, just widen-ing the many relevant findings of experimental economics with human agents, and checking their robustness against alternative controlled agents' behaviour.

The first experiment with programmed agents (beyond experimental economics) by Gode and Sunder [5] was a big surprise. They confirmed that institutions matter. To the extreme that in a CDA price convergence and allocative efficiency was achieved, even with zero intelligent (poorly instructed but perceptive) agents. Spontaneous order arises in the CDA thus confirming Hayek and A. Smith conjectures.

Since Gode and Sunder [5] several studies have examined CDA with various com-puterized bidding agents. Cliff and Bruten [2] designed the "zero intelligence plus" (ZIP) agents to demonstrate that institutions matters and so does intelligence. They employed an elementary form of machine learning to explore the minimum degree of agent intelligence required to reach market equilibrium in a simple version of the CDA. Preist and van Tol [10] used different heuristics for determining target profit margins in ZIP agents to faster achieve market equilibrium. Gjerstad and Dickhaut [4] proposed an agent (GD) who places the bid or offer that maximizes the expected surplus (calculated as the product of the gain and the probability for the bid or offer to be accepted). Das *et al.* [3] made improvements on the GD agents and described a series of laboratory experiments that, for the first time, allow human subjects to interact with two types of software bidding agents (ZIP and GD). They found that agents consistently obtain significantly larger gains from trade than their human counterparts. Grossklags and Schmidt [6] found that the information on the existence of software agents in the market environment results in more efficient behaviour of human traders.

We focus on the interactions between software bidding strategies. In this sense, Tesauro and Das [14] tested agent performance in homogeneous populations and in two heterogeneous settings: (1) a single agent of one type competes against an otherwise homogeneous populations of a different type; (2) two types of agents compete in a market where one has a counterpart of the other type.

A typical approach to evaluate strategies in heterogeneous populations has been to establish a tournament, like the Santa Fe Double Auction (Rust *et al.* [11]) or the Trading Agent Competition (TAC) (Wellman *et al.* [16]). Walsh *et al.* [15] said that the tournament-play is one trajectory through the space of possible interactions and the question of which strategy is the best is often not the most appropriate given that a mix of strategies may constitute an equilibrium. They proposed a method for analyzing the interaction among heterogeneous heuristic strategies. They used the replicator dynamics formalism to model the evolution in the strategy space basing in a created heuristic payoff table.

Our work differs from prior CDA studies on the interactions between bidding strategies (Tesauro and Das [14], Walsh *et al.* [15]) in three ways. (1) These works assume fixed strategies for an agent over time. We allow the soft-agents to learn switching between three alternative strategies. (2) We examine the behaviour of strategic traders under general supply and demand schedules (asymmetric environments) not only under symmetric environments. (3) Our approach is Agent Computational Economics and the mix of strategies equilibrium emerges from the bottom-up versus the top-down approach of Walsh *et al.* [15].

Since auction design is a question of scarcity and wants, we may give back to Agent Theory and Artificial Intelligence a nice heritage from the experimental economics evidence: treat management engineering problems as auctions (market oriented programming, Boutilier *et al.* [1]). At INSISOC (www.insisoc.org) we are doing research in some fundamental areas in production and management with this new focus.

3.2 Intelligent Agents in the CDA Market. The Model

We describe our model in terms of the essential dimensions of any market experiment (IxExA).

3.2.1 The Institution

The institution is a Continuous Double Auction (CDA). The CDA imposes no restrictions on the sequencing of messages. Any trader can send a message at any time during the trading period. We consider a CDA with a bid-offer

spread reduction. The new bid/offer has to provide better terms than previous
outstanding bids/offers.

3.2.2 The Environment

Each trader is either a seller or a buyer. The assumption of fixed roles conforms
to extensive prior studies of the CDA, including experiments involving human
subjects and automated bidding agents. Each agent is endowed with a finite
number of units. Seller i has ni units to trade and he has a vector of marginal
costs $(MaC_{i1}, MaC_{i2}, \ldots, MaC_{in_i})$ for the corresponding units. Here MaC_{i1}
is the marginal cost to seller I of the first unit, MaC_{i2} is the cost of the second
unit, and so on. Buyer j has n_j units to trade and he has a vector of reserve
prices $(RP_{j1}, RP_{j2}, \ldots, RP_{jm_j})$ for the corresponding units. Here RP_{j1} is the
reserve price to seller I of the first unit, RP_{j2} is the reserve price of the second
unit, and so on. These valuations are private.

Our model does not have environmental restrictions and it allows us to
simulate any environment in terms of the number of traders, their units and
the valuations of each trader. So we can simulate competitive as well as market power environments, both with symmetric or asymmetric supply and demand curves. We consider that an environment is symmetric if the supply
and demand curves have opposite signs but equal magnitudes. Otherwise, we
consider that the environment is asymmetric.

3.2.3 Agents' Behaviour

In CDA markets traders face three non-trivial decisions: How much should
they bid or offer? When should they place a bid or an offer? And when should
they accept an outstanding order? Each agent type: ZIP, GD and K (defined
below) corresponds to particular values for these decisions.

Each ZI-Plus agent (Cliff and Bruten [2]) has a mark up μ that determines
the price at which he is willing to buy or sell in adaptive way. The agents
learn to modify the profit margin over the auction using the information about
the last market activity. For example, the profit margin of a buyer is

$$\mu = 1 - \frac{howMuchBid_{t-1} + \Delta_t}{ReservePrice}, \tag{3.1}$$

where Δ_t is calculated using the individual trader's learning rate (β), the
momentum learning coefficient (γ) and the difference between the target bid
and the bid in the last round.

The GD agent is a more sophisticated one (Gjerstad and Dickhaut [4]).
Each seller chooses the bid that maximizes his expected surplus, defined as the
product of the gain from trade and the probability for an offer to be accepted.
GD agents modify this probability using the history HM of the recent market
activity (the bids and offers leading to the last M traders: ABL accepted bid

less than b, AL accepted bid and offer less than b, RBG rejected bid greater than b, etc.) to calculate a belief function. Interpolation is used for prices at which no orders or traders are registered in HM. For example, the belief function of a buyer is:

$$\hat{q}(b) = \frac{ABL(b) + AL(b)}{ABL(b) + AL(b) + RBG(b)}. \tag{3.2}$$

The Kaplan (K) agent is the third type agent we consider. It was the winner in the tournament of Santa Fe Institute in 1993 (Rust *et al.* [11]). The basic idea behind the Kaplan strategy is: "wait in the background and let others negotiate. When an order is interesting, accept it". K agents must be parasitic on the intelligent agents to trade and to obtain profit. If all traders in the market are K agents no trade will take place.

In our model, we consider one more decision: Which strategy should they choose to obtain higher profit? Each agent chooses a strategy from a set of three alternatives (GD, K and ZIP). To take this decision each trader only knows their own reservation prices and the information generated in the market, but he doesn't know the bidding strategy of the other agents or the profit achieved by them.

Each agent learns to change his strategy looking for the best bidding strategy in the following way: An agent will consider to change his strategy if the profit is less than the profit from the previous period. The agent will actually change his strategy if he believes that he could have reached higher profit following an alternative strategy.

In table 3.1 we show how a buyer forms his beliefs. The buyer compares if the bid of an alternative strategy (BA) could have been lower or greater than the realized bids or accepted offers under the current strategy. Following this comparison, he may conclude:

- If BA is lower than the realized bid and greater than the minimum transaction price for that period, a buyer will consider that the BA would have been accepted and he could have obtained greater profits. But if it is lower than the minimum transaction price, a buyer will assume that no seller would have accepted the BA.
- If BA is greater than the realized bid, a buyer could have obtained lower profits.

Table 3.1. Buyers beliefs

	Realized bids	Accepted offers
Lower	Greater profits (if it is greater than the minimum transaction price) No profit (if it is lower than the minimum transaction price)	No profit
Greater	Lower profits	The same profits

- If BA is lower than the accepted offer, a buyer would have rejected the offer and he could have obtained no profit.
- If BA is greater than the accepted offer, a buyer could have obtained the same profits whatever the value of the bid was.

Following this strategy choice decision, the resulting population will have a particular proportion of the three types of software bidding agents (ZIP, GD and K). To have a graphical idea of the dynamics of this proportion, we will represent the strategy space by a two dimensional simplex grid with vertices corresponding to the pure strategies (all ZIP, all GD or all K). We draw three regions to represent the populations that have a dominant strategy when more than 33% of the agents use the strategy.

3.3 The Experiments

Twenty agents (10 buyers and 10 sellers) compete in the market. For a game with 20 agents, each one with 3 strategies (ZIP, GD and K), there will be 231 possible populations. For each one of these possible populations, 30 simulations are run. Each agent is given a list of ten limit prices (valuations). To isolate the effects of behaviour under symmetric or asymmetric environments and to prevent some agents having relative initial advantage, the agents' valuations are the same for all the agents that are in the same side of the market.

The simulations can be represented in six scenarios, that accommodate three different environments and two kinds of agent strategies (fixed strategies and changes on strategies. See table 3.2).

Table 3.2. Simulated environments. E1F stands for symmetric environment and fixed strategy, E2C for perfectly elastic supply and changes in the strategies, and so on

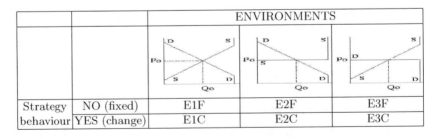

		ENVIRONMENTS		
Strategy	NO (fixed)	E1F	E2F	E3F
behaviour	YES (change)	E1C	E2C	E3C

The scenarios of fixed strategies (E1F, E2F and E3F) allow us to evaluate the effects of the changes in the environment. For a given environment we can check the gain of learning to choose the strategy.

3.4 Some Results and Discussion

3.4.1 Fixed Learning Strategies During the CDA

Price Behaviour in Homogeneous Populations

We first compare the agent performance in homogeneous populations under different environments. These elemental experiments are a starting point and give us an idea of the convergence patterns of each type of software bidding agents (ZIP, GD and K). Table 3.3 presents the time series of transaction prices over three periods (100 rounds per period).

If all traders in the market are K agents, no trade will take place. In the other cases (all ZIP or all GD) the transaction prices converge to the competitive equilibrium price from above (below) when the supply (demand) is more elastic than demand (supply). Comparing the first column with the second one, we observe that GD agents take less time than ZIP agents both to learn and to exchange under any environment. In homogeneous GD populations the transaction are made in the first rounds of each period and the prices are very close to the competitive equilibrium price.

Table 3.3. Transaction prices in homogeneous populations

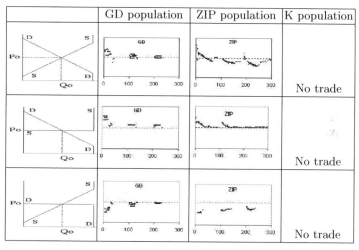

Price Behaviour in Heterogeneous Populations Where More than 50% are K agents

In these cases the convergence to competitive equilibrium is not achieved (see table 3.4). K agents extract surplus from the other side of the market. Consequently the transaction prices move towards the other side of the market.

Table 3.4. Transaction prices and profits: 50% K agents and 50 % GD agents

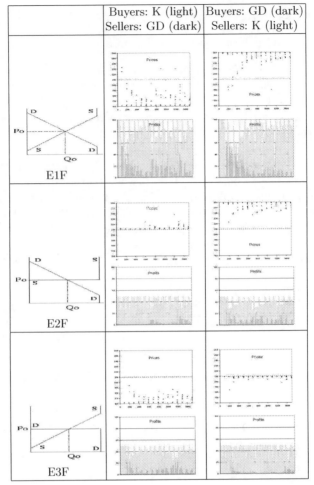

Under symmetric environments (E1F), if all buyers are K agents the transactions prices are in the sellers side. While if all sellers are K agents the transaction price are in the buyers side. This effect is bigger under asymmetric
environments.

If supply is perfectly elastic (E2F) and all sellers are K agents, their profit
should theoretically be cero because their valuations are equal to the competitive equilibrium price. But this does not happen. As in the symmetric case
above, they steal all the buyers surplus. Accordingly the profits achieved by
the K agents (light bars) are greater than the profits of GD agents (dark bar).
The same results hold for K agents against ZIP agents. No matter in which
side of the market they are or what is the elasticity of their side.

Market Efficiency

Market efficiency is achieved in almost all the space of possible interactions. Only in heterogeneous population where more than 50% are K agents, the efficiency goes down. This confirms previous results for symmetric environment (Posada *et al.* [9]). But under asymmetric environments we get higher volatility.

3.4.2 Changing the Strategy During the CDA

Market Efficiency

Market efficiency is always achieved when the agent learns how to change his strategy. This happens for any initial proportion of the types of agents and for any kind of environment.

Price Behaviour

Convergence to the competitive price is achieved after some previous learning under any initial population and any kind of environment (see Table 3.5).

Strategy Space Evolution

The profit bar charts of table 3.5 indicate that, the following patterns emerge. GD agents settle in the more elastic side of the market. The Kaplan agents can free ride only in the case that they all are in one side of the market and no change of strategies is allowed. This is further confirmed in the simplex strategy space grid (see table 3.6).

In symmetric environments(EIC) there is an "attractor" zone, where no strategy seems to dominate but no clear Nash equilibrium comes up (point A) If it is asymmetric there is an "attractor" zone, point B, (8 K, 10 GD, 2 ZIP) no matter which side is the elastic one. In both cases (E2C, E3C) the GD agents settle in the elastic side.

3.5 Conclusions and Further Research

We have extended in several ways related works following Gode and Sunder [5] seminal contribution in experimental economics with programmed agents: beyond experimental economics.

Price dynamics for a CDA is obtained for non symmetric environments and populations of heterogeneous agent behaviour. We allow the agents to change their strategies endogenously so that we can trace the patterns of the emerging proportion of agents' behaviour not only with symmetric environment as in Posada *et al.* [9] but also with asymmetric environments.

For a complete range of the possible settings (ExA), the results confirm that:

Table 3.5. Transaction prices and profits when the initial population is 50% are K agents and 50% GD agents

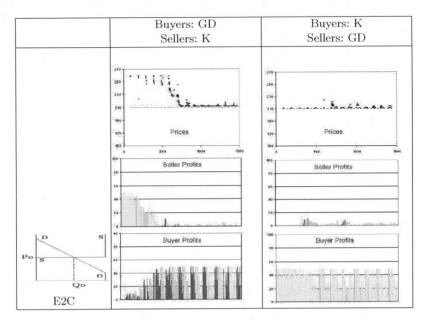

1. The quality of price convergence and allocative efficiency, depend on alternative degrees of agents intelligence.
2. Price dynamics and agents surplus depend on the proportion of the types of in-telligent agents: Kaplan Zero intelligent, G.D.
3. It also matters whether the environment is symmetric or not. Nevertheless convergence is achieved if we allow the agents to change their strategy. There is not Nash equilibriums in the strategy proportions, but under asymmetric environments the GD strategy becomes dominant.

These results clarify the scope and limitations of A. Smith and Hayek's observations on emerging market order for a wide range of variations of (ExA) in the con-text of a particular but important Institution (I), the Continuous Double Auction. Institutions matter and so does intelligence as far as the rate of convergence and agents surplus is concerned.

The approach can be easily extended, in the same vein, to study the equivalence of alternative Institutions for fixed (ExA) settings. One can easily extend the approach to include speculation agents used to trade in B2B where demand and sup-ply fluctuate from period to period (Li and Smith [7]). A more complex issue will be to consider a multi-product auction and the possibility of agents trading in both sides of the markets at the same time.

These results are of importance in the design and functioning of auctions in the real world and in the applications of auction theory to many problems

Table 3.6. Strategy space evolution

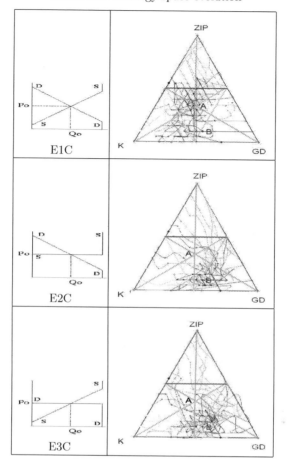

in management and production far beyond market design (market oriented programming).

Acknowledgements

This work has received financial support from the Spanish MEC, n° 2005-05676. We wish to thank three anonymous referees for their helpful comments.

References

[1] Boutilier, C., Shoham, Y. and Wellman, M.P. (1997) Economic principles of multiagent systems (Editorial) (). Artificial Intelligence, 94: 1-6.

[2] Cliff, D. and Bruten, J. (1997) Zero is not enough: On the lower limit of agent intelligence for continuous double auction markets. HP-1997-141, Hewlett Packard Laboratories, Bristol, England.

[3] Das, R., Hanson, J., Kephart, J. and Tesauro G (2001) Agent-Human Interactions in the Continuous Double Auction. In Proceedings of the International Joint Confer-ences on Artificial Intelligence.

[4] Gjerstad, S. and Dickhaut, J. (1998) Price formation in double auctions. Games and Economic Behavior 22: 1-29.

[5] Gode, D. and Sunder, S. (1993) Allocative efficiency of market with zero-intelligent traders: Market as a partial substitute for individual rationality. J. of Political Economy 101:.

[6] Grossklags, J. and Schmidt C. (2003) Interaction of Human and Artificial Agents on Double Auction Markets. Simulations and Laboratory Experiments. In Proceedings of the 3rd International Workshop on Computational Intelligence in Economics and Finance.

[7] Li, L. and Smith, S. (2004) Speculation Agents for Dynamic, Multiperiod Continuous Double Auctions in B2B Exchanges. In Proceedings 37th Hawaii International Conference on System Sciences.

[8] López, A., Hernández, C. and Pajares, J. (2002) Towards a new experimental socio-economics. Complex behaviour in bargaining. Journal of Socio-Economics 31: 423-429.

[9] Posada, M., Hernández, C. and López, A. (2005) Learning in a Continuous Double Auction Market. In: Mathieu P., Beaufils B. and Brandouy O. (Eds), Artificial Economics - Lecture Notes in Economics and Mathematical Systems 564. Springer-Verlag, Berlin, 2005.

[10] Preist, C. and van Tol, M. (1998) Adaptive agents in a persistent shout double auction. In Proceedings of the First International Conference on Information and Computation Economies, pages 11–18. ACM Press.

[11] Rust, J., Miller, J. and Palmer, R. (1993) Behavior of trading automata in computerized double auctions. In: Friedman and Rust (eds.), The double auction markets: Institutions, theories and evidence. Addison-Wesley.

[12] Smith, V. (1989) Theory, Experiment and Economics. J. of economic Perspectives. Winter. 783-801.

[13] Sunder, S. (2005) Markets as Artifacts: Aggregate Efficiency from Zero-Intelligence Traders. In Models of a Man. Essays in Memory of H. Simon. Augier and March edit. MIT Press: 501-519.

[14] Tesauro, G. and Das, R. (2001) High performance bidding agents for the continuous double auction. In: Third ACM Conference on Electronic Commerce.

[15] Walsh, W., Das, G., Tesauro, G. and Kephart, J. (2002) Analyzing complex strategic interactions in multiagent games. In: 8th C. on Artificial Intelligence, Canada.

[16] Wellman, M., Wurman, P., O'Malley, K., Bangera, R., Lin, S., Reeves, D. and Walsh, W. (2001) Designing the market game for a trading agent competition. IEEE Inter-net Computing 5(2):43-51.

Part II

Market Efficiency and the Role of Speculation

4

A Broad-Spectrum Computational Approach for Market Efficiency

Olivier Brandoy[1] and Philippe Mathieu[2]

[1] LEM, UMR CNRS-USTL 8179, France olivier.brandouy@univ-lille1.fr
[2] LIFL, UMR CNRS-USTL 8022, France philippe.mathieu@lifl.fr[*]

4.1 Introduction

The Efficient Market Hypothesis (EMH) is one of the most investigated questions in Finance. Nevertheless, it is still a puzzle, despite the enormous amount of research it has provoked. For instance, many recent results have shadowed the well-established belief that market cannot be outperformed in the long run (Detry and Gregoire [2]).

One other reason is that persistent market anomalies cannot be easily explained in this theoretical framework Shiller [11]. Additionally, one can also consider that some talented hedge-fund managers (like Jim Simons) keep earning excess risk-adjusted rates of returns since years. Nevertheless, there is no consensus on this last point today Malkiel [7].

Many versions of the EMH have been proposed since the founding works of Samuelson [10]. We concentrate in this paper on the weak form of efficiency Fama [3]: *"past informations are useless to predict future price changes"*. We, therefore, focus on the efficacity of simple technical trading rules, following Jensen and Benington [6] or more recently Brock et al. [1]. An extensive survey for this issue is proposed in Park and Irwin [8].

Nevertheless, we depart from previous works in many ways: we first have a large population of technical, virtual agents (more than 260.000) exploiting real-world data to manage a financial portfolio as chartists or technical traders would do. Very few researches have used such a large amount of calculus to examine the EMH. Our experimental design allows for agents selection based on past absolute performance, as well as consistency of performance. We take into account the data-snooping risk, which is an unavoidable problem in such broad-spectrum researches, using a rigorous *Bootstrap Reality Check* (BRC) procedure [12].

[*] This work has received a grant from European Community – FEDER – and *"Region Nord-Pas de Calais"* – CPER TAC –.

While market inefficiencies, after including transaction costs, cannot clearly be successfully exploited, our experiments present troubling outcomes like persistent (but not statistically significant at commonly admitted levels) over performance, inviting close re-consideration of the weak-form EMH.

This research is organized as follows: section 4.2 presents our multi-agent system (MAS) and experimental design, section 4.3 is dedicated to our results and section 4.4 concludes this research.

4.2 Methodology

The methodological section gives the main features of our experimental design, including the global MAS architecture, descriptions of the agents, and the statistical procedure aimed at detecting potential market inefficiencies.

4.2.1 MAS Architecture

In this experiment, agents represent virtual investors trading a single financial commodity called *"a tracker"*. As it is generally admitted [13], the agents' fundamental characteristics in this study are an idiosyncratic decision-making process, autonomy and reactivity to contextual changes. Our MAS is based on a three-stage architecture (see figure 4.2); at each stage, one can consider a particular kind of agents with specific aims or logic:

First stage: *Strategic Agents* are micro-agents always playing the same basic strategy through the entire simulation. Those basic strategies are known in the financial community as "technical trading rules". For instance, a *Strategic Agent: "5-days moving average"* cannot process any other operation and has to decide whether to trade or not on the basis of a single rule.

Second stage: *Family Agents* are general agents defining all the formal characteristics used in the instantiation of each *Strategic Agent*. Each *Family Agent* also has to perform a ranking between each of his "children" at each time step. The *Family Agent*, thus, has the capacity to select the most successful individuals among the *Strategic Agents*. For instance, the *Family Agent "Rectangle"* combines four parameters (n, m, p, and s, see figure 4.1).

Third Stage: *Meta Agents* are able to mimic the behavior of various *Strategic Agents* according to the circumstances and the ranking given by the relevant *Family Agent*. For instance, a *Meta Agent* based on the 2-uple $\{Momentum, Triangle\}$ will choose and mimic various instances of those *Family Agents*, after considering some signals. For instance, it can begin with replicating a *Strategic Agent: "5-days momentum"* and then keep on going with this for eight days, then switch to replicate a *Strategic Agent: "3-days triangle"* for the next six iterations and so on... We do not develop this point in this article, nor do we report results concerning this category of agents.

To get it clearer, let's consider one *Family Agent: "Periodical Trader"*. This agent buys and sells the trackers at fixed intervals. It is similar to speculators buying on Mondays and selling on Fridays. This agent has at least two parameters coding the dates on which it will buy and sell the trackers. If it decides to generate all possible *Strategic Agents* using all possible delays between 1 and 100, 10.000 "children" will be processed. In this study, we have 10 *Family Agents* generating more than 260.000 *Strategic Agents*. One can imagine that the number of *Meta Agents* is, therefore, really huge and, despite computer power or parallel computing facilities, cannot be investigated exhaustively.

4.2.2 Agent's Design

Agent's design is specified in terms of the decision making process and operations allowed in the market.

Agents Population:

As has been presented previously, we have implemented a large population of heterogeneous agents (267.069 agents, see Table 4.1).

Among these strategic agents, 264.117 (98.89%) are never bankrupted during the whole process.

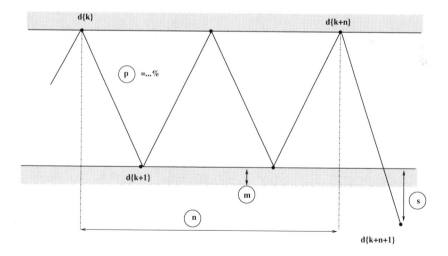

Fig. 4.1. Family Agent "Rectangle"

Table 4.1. Agents Population

Family Agent	Num. Strategic Agents
Periodical $\{n, m\}$	250.000
Indicator $\{n, m, p\}$	1.470
Rectangle $\{n, m, p, s\}$	7290
Triangle $\{n, m\}$	1.547
Variation $\{n, m, p\}$	2.000
Momentum $\{n, m\}$	220
Moving Average $\{n\}$	200
Weighted Moving Average $\{n\}$	200
RSI $\{n, p\}$	4.141
Buy & Hold	1
Total	267.069

Allowed Operations and Behavior:

Each agent is allowed to trade n trackers ($n \in R^+$), that is, one financial commodity replicating exactly one market index (like CAC40, Dow-Jones or Nikkei). If it has not decided to hold such commodities, the agent holds cash. Therefore, each agent is in one of these situations:

- it possesses a number of trackers > 0; in this situation we say that the agent is *"in the market"*. Its wealth fluctuates along with the market.
- it does not have any tracker or fraction of a tracker, all its wealth been converted into cash; the agent is *"out-of the market"* and its wealth is

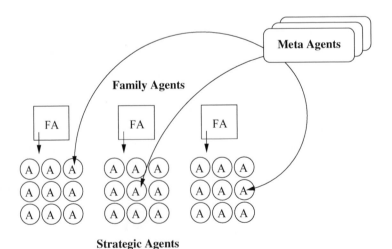

Fig. 4.2. Multi-Agent System design

stable. There is no risk-free asset paying a low interest rate available is our simulations.

At each time step, agents receive new information and have to decide if it is worth staying in the market or getting out: they follow systematically the signals given by their technical rules. For instance, a *Strategic Agent* designed as a *"moving average 5, 5"* analyzes at each iteration if the five days moving average of past prices has crossed the price process in the top-down direction, which correspond to a "sell-signal" (*resp.* bottom-up, "buy-signal"). In this situation, if the gap between the five days moving average and the price is greater than 5%, it will "sell" (resp. "buy"). If the gap is under 5%, it will keep its portfolio unchanged. Each *Strategic Agent* follows the same kind of behavior with various charts or technical rules. Nevertheless, one has to notice than one singular agent follows systematically a *"Buy & Hold"* strategy (B&H), that is, it enters the market at *time* = 0 (buys one tracker) and lets the situation remain unchanged until the end of the simulation. This agent is our "benchmark" agent in terms of risk and return and stands for a "passive investor".

Theoretically, no one can outperform this agent when considering the risk-adjusted performance in the long run, assuming the EMH holds. In other words, despite it is obvious than anyone can construct a portfolio or adopt a strategy that will outperform the B&H strategy, however, this assumes a higher risk level for the investor and, generally speaking, cannot be qualified as an outstanding behavior.

Each agent is endowed with the same amount of cash at the beginning of the simulations. If an agent looses all its endowment during an experiment, since borrowing is not allowed, it is withdrawn from the market.

In the simulations, agents are considered as *"price takers"*, that is, their behavior has no effect on the price of the asset they trade. This is a very commonly accepted hypothesis in finance, whereas it can be debated in MAS dealing with artificial stock markets. Trading is subjected to transaction costs at a 0.5% level.

	Real–Worl Universe (RWU)	Artificial–World Universe (AWU)
Subperiod 1 *in–sample selection*	CAC 40 01-1988 : 07-1996	i.i.d Random–Walk
Subperiod 2 *out–of sample test*	CAC 40 08-1996 : 04-2005	i.i.d Random–Walk

Fig. 4.3. Experimental design

The simulations are based on real daily data from the Euronext Paris Stock Exchange between 1988 and 2005. The traded tracker perfectly replicates CAC40 index. Agents have access only to past values of this tracker and the information they receive at each time-step is the price of the tracker corresponding to the current iteration (no agent is "cheating" and none behaves like knowing what the "future" will be).

4.2.3 Organization of the Simulations

Our experimental design is organized in two steps on two *"universes"* (see Fig. 4.3):

1. *"Universes"* are sets of data used to perform the simulations. Simulations are parallelized over the universes, each of them being useful for understanding what happens in the other.
    a) Real-World Universe: consists in historical CAC40 observations (see fig. 4.4), split into two subsamples, RWU_1 (01/1988-07/1996) and RWU_2 (08/1996-04/2005).
    b) Artificial Universe: consists in computer-generated data using an *iid* random-walk process[3]. This universe is also split into two subsamples AU_1 and AU_2 and includes the same number of observations as in RWU_1 and RWU_2. These sets of data are intended to provide a universe in which it is actually impossible to outperform the market since it is artificially generated (assuming the random generator is good enough).
2. Over these universes, the simulations are organized as follows:
    a) Step 1, *"in-sample selection"*: is the selection of the best performing agents, compared to the benchmark agent. This test consists of 10 simulations based on random subsamples picked in RWU_1 and AU_1. These subsamples will be called *windows*. At the beginning of each simulation $(t = 0)$, *Family Agents* create *Strategic Agents*. Then *Strategic Agents* begin to compete against the *Buy & Hold Agent*. Then *Family Agents* rank their respective sub-populations of *Strategic Agents*, comparing their performances with that of the benchmark agent. Once the 10 simulations have been processed, *Strategic Agents* that have out-performed the benchmark at least in 50% of the simulations are selected.

    Performance is always appreciated in terms of risk-return: a *Strategic Agent* outperforms the *Buy & Hold Agent* if and only if it achieves a more than proportional return considering the risk it has been exposed to during the simulation. "Risk" is calculated as the standard deviation of the agent's portfolio, "return" being the average rate of growth of its wealth.

---

[3] $p_t = p_{t-1} + \varepsilon_t$ with $\varepsilon_t \rightarrow N(\mu, \sigma)$, μ and σ being chosen to fit as closely as possible the corresponding parameters in RWU_1 and RWU_2

b) Step 2, *"out-of sample tests"*: consists of generalization of the first stage using the relevant second subsamples (*i.e. Strategic Agents* having out-performed the benchmark at least in 75% of the simulations.

4.2.4 How Do We Decide if a *Strategic Agent* Outperforms the *B&H* Agent?

Three performance indices are calculated providing information on risks and returns of the *Strategic-Agents*:

1. Return: $\overline{r_i}$ is the daily return earned by each agent i, for *windows* $t = [1, n]$, using the following formula:

$$\overline{r_i} = [Port_{i,t=n} - Port_{i,t=1}]^{1/n} \qquad (4.1)$$

In equation 4.1, $Port_{i,t}$ is the agent's i portfolio on date "t".

2. Risk: is calculated as the standard deviation of $r_{i,t}$ on each corresponding *window*.

3. Synthetic Index: combines the preceding indices and provides an aggregated measure for the absolute performance of one specific Agent i:

$$SI_i = r_i/\sigma_i \qquad (4.2)$$

One can notice the Synthetic Index reported in equation 4.2 is very similar to a Sharpe Index.

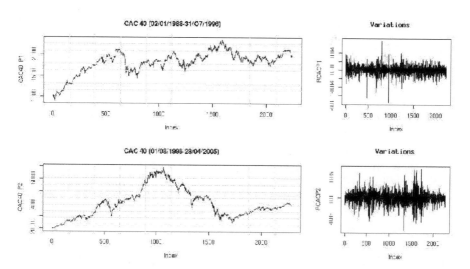

Fig. 4.4. RWU_1 and RWU_2 data – level / variations –

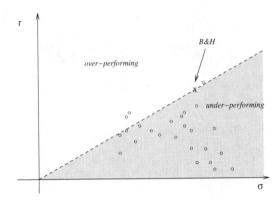

Fig. 4.5. Outperforming Strategic Agents in the Risk-Return Space

This set of indices is systematically evaluated for pairs of agents on each *window*. These pairs of agents are always a combination of one *Strategic Agent* and the *B&H* Agent. This procedure allows us to place *Strategic Agents* in a risk-return space for subsamples of observations. Assuming we have 10 *windows*, we thus will have to consider 10 risk-return spaces. In these spaces, outperforming *Strategic Agents* are placed in a part of the plan above the line crossing the origin and reaching one point representing the performance of the *B&H* Agent (see Fig. 4.5).

4.2.5 The Data Snooping Issue

Although the process presented in 4.2.3 might appear to be very harsh, it is clearly not sufficient to "prove", if at all it is possible, that any persistent, abnormal over-efficiency of some specific agents really occurs. Since we investigate the performance of a very large set of agents, we must consider the *data-snooping* problem.

To give an illustration of "data-snooping", let's consider the following example (derivated from Jensen and Cohen [4]): suppose you would like to hire someone having *abilities* to predict the next movements in a particular stock exchange. Obviously, the person to be hired would have to perform this task better than merely taking chances. To select a good candidate, you propose the following test: *"predict the next 14 fluctuations of the stock exchange in the following terms: 0 if the market closes up, 1 if it closes down".* In other words, each candidate would have to propose a 14 characters-long string like 00101110010101. Suppose now you decide to hire someone providing at least a 75% rate of correct predictions (at least 11 good answers). The probability for someone to succeed here only by chance is very weak:

$$\sum_{i=11}^{14} C_{14}^i 0.5^{14} \simeq 0.02869$$

In other words, someone without any skill to predict these fluctuations would roughly have a 97% chance to fail. Suppose now 10 candidates face this challenge, none of them having any particular ability to predict the stock exchange, then the probability that *"at least one of them would succeed"* is sufficiently large to warrant careful examination of the successful candidate:

$$1 - (0.9713)^{10} \simeq 0.2525$$

Basically, if you increase the number of applicants to a certain point, you will probably hire someone passing the test[4] but nothing proves that this person has performed better than merely taking chances. This problem, known as the "data-snooping" bias, has been recognized very early in financial research, where data-mining has a long tradition [5][5]. One way to mitigate this issue is to apply a procedure called *Bootstrap Reality Check* (BRC) proposed by White (2000). In this research, BRC is intended to decide whether or not the selected agents, at the end of our experimental procedure, have positively out-performed the benchmark or not, that is, if they have out-performed the market exploiting weak-form inefficiencies or if this result must be attributed to chance.

Bootstrap Really Check (BRC) Procedure

BRC consists of testing the following null hypothesis: *"H_0: the best Strategic Agent does not outperform the B&H agent.*

Let's note θ_k one specific performance index for the k-th *Strategic-Agent* in a set of M agents,

$$H_0 : \max_{k=1...M} E(\theta_k) \leq 0 \qquad (4.3)$$

This performance is calculated over n subsamples (n=200) taken from RWU_2 or AU_2.
In this research, θ_k is:

$$\overline{\theta_k} = 1/n \sum_{T=1}^{n} (SI_{T,i} - SI_{T,B\&H}) \qquad (4.4)$$

---

[4] with 100 candidates, the probability than none of them succeed is around 5%

[5] *"Let us [...] assume that we have access to a large computer and a body of security price data. Now, if we begin to test various mechanical trading rules with enough variants, we will eventually find one or more which would have yield profits [...] superior to a buy and hold policy [...] We cannot be certain that [...] results did not arise from mere chance"*

In equation 4.4, T is a specific *window*. In other words, we focus on an average over performance, appreciated with the *Synthetic AI Index* (see 4.2.4), over all the n considered *windows* for the selected agents. The best agent will therefore have an estimated performance as follows:

$$\overline{V} = \max_{i=1...M} (\sqrt{M} \times \overline{\theta_i}) \tag{4.5}$$

We then generate B bootstrapped series using a process described by Politis and Romano [9]. Over those bootstrap series, we estimate again the whole set of performance indices θ_i. To distinguish these indices from those coming from the initial set of data, we note them $\theta_{b,i}$ (b being the *b-th* bootstrap series). The *p-value* for the Null is therefore:

$$p = \sum_{b=1}^{B} \frac{Z_b}{B} \tag{4.6}$$

with

$$Z_b = \left\{ \begin{array}{l} 1 \text{ if } \overline{V_b^*} > \overline{V} \\ 0 \text{ else} \end{array} \right| \tag{4.7}$$

4.3 Results

The simulations have been conducted over two different universes, as explained previously (see 4.2.3). Our results are presented successively for each universe.

4.3.1 Do Strategic Agents Behave Well in the Artificial Universe?

By construction, the Artificial Universe does not "hide" any useful information on date t allowing to predict what will probably happen on date $t+1$. Thus, these data perfectly replicate the behavior of an efficient market index. No *Strategic Agent* should be able, in this specific, virtual context to pass the filters we have programmed. Only chance could explain such an improbable success. Table 4.2 presents the best agents after each simulation step.

After Step 2, *none* of the Strategic Agents could be selected with the required 75% success rate. Table 4.2 shows the number of *Strategic Agents* outperforming the *B&H* agent in at-least 50% of the windows.

This leads us to consider the following explanation for this series of simulations:

- Our simulation process is sufficiently harsh and proves its efficiency in selecting good candidates: when no structure is hidden in a time series, no agent can outperform a basic *B&H* behavior.
- Our Artificial World does not reflect properly the real-world data (mainly because we have designed it as an *i.i.d.* process), and more complex dynamics in the Artificial World might have given a different result (ARCH process as example).

4.3.2 Does *RWU* Exhibit Inefficiencies?

The whole set of results concerning the two-step selection process exposed in
4.2.3 is reported in Table 4.3.

In-sample Selection: RWU_1.

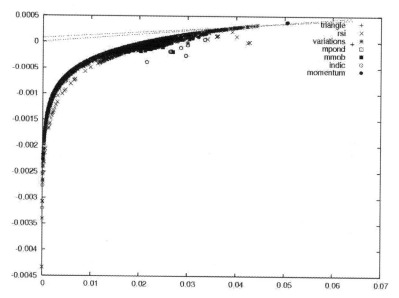

Fig. 4.6. *Strategic Agents* in the risk-return space for one window of RWU_1

Table 4.2. Agents "outperforming" the Artificial Index

Family Agent	Num. *after Step 1*	Num. *after Step 2* 50% selection rate
Periodical	13368	3
Indicator	13	–
Rectangle	286	2
Triangle	20	–
Variation	28	1
Momentum	13	–
Moving Average	14	–
Weighted Moving Average	3	–
RSI	76	2
Total	13.821	8

Here, many *Strategic Agents* (6.057) outperform the benchmark agent in more than 50% of the cases. Each *Family Agent* has at least one of its children selected at the end of this simulation step. The major part comes from the *Periodical Family Agent* which is *"as expected"* but not significant since this family does not rely on classical technical signals "revealing" a "pattern". Fig. 4.6 shows, for one window in RWU_1, *Strategic Agents* from various *Families* in the risk-return space. The characteristic concave shape of the plot can be explained by the weigh of transaction costs that penalize the most active agents.

The selection rate over the initial population is between 0.34% and 11.5%, which is low, but "as expected". One has to keep in mind that our procedure involves a very large number of agents; it is perfectly normal that some of them seem to perform well at this initial stage. Fig. 4.7 shows the behavior of some interesting agents in terms of level of portfolio.

Out-of Sample Tests: RWU_2 and BRC

After the second selection process, only 19 *Strategic Agents* have out-performed the benchmark agent. They come from just two *Family Agents*: *Rectangle Trader* and *Variation Trader*. Some of them have outperformed the benchmark agent in each of the 200 simulations.

Clearly, the proportion of "good candidates" at the end of this out-of-sample test is very low. This is not surprising since modern stock markets are obviously not inefficient.

Fig. 4.8 shows the behavior of two very interesting *Strategic Agents*, *variation 7, 10, 3* and *variation 7, 10, 13*. The first one obtains a 100% score over 200 random windows in RWU_2 while the second one obtains a 76% score. Lines show the portfolio of these agents and the *B&H*'s for the specific window (01/2003-04/2005).

The next step in the analysis is to verify if this result can provide a kind of basis to reject the weak-form EMH. Thus, we have applied carefully White's *Reality Check* over 500 bootstrap series to control potential spurious results. The procedure leads to consider again the whole set of 6.057 *Strategic Agents* passing the first selection step.

Although the simulations seem to be very harsh in terms of selectivity for "good candidates", we cannot reject the null hypothesis: *"The best Strategic Agents cannot out-perform the Buy & Hold Agent"* at ordinary p-values ($p\text{-}value=28.2\%$). Therefore, we cannot reject the initial weak-form EMH and cannot report evident market inefficiencies for data *with basic Strategic Agents* using simple trading rules. This result seems, therefore, to be a strong support for the weak-form efficiency of the French Market.

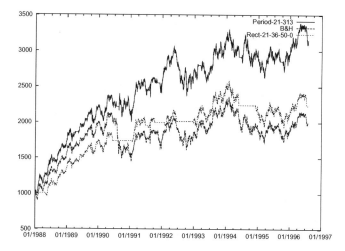

Fig. 4.7. Examples of "good *Strategic Agents*" on RWU_1

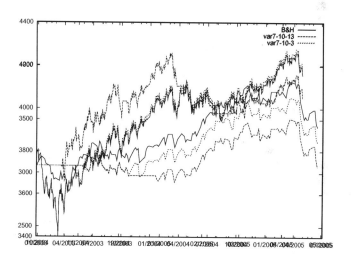

Fig. 4.8. Portfolios of two *Strategic Agents* and *B&H Agent*

4.4 Concluding Remarks

In this research, we show that technical traders cannot outperform a simple *Buy and Hold* Agent on Paris Euronext Stock Exchange. This result is derived from the following observations: our MAS can select some (apparently) very robust agents, producing very good risk-return scores after a harsh filtering process.

Table 4.3. Simulations: Real-World Universe

Family Agent	Stage 1		Stage 2	
	Number	% / Initial Population	Number	% Remaining Population
Periodical	5.484	2.19%	–	–
Indicator	5	0.34%	–	–
Rectangle	367	5.03%	12	3.27%
Triangle	74	4.78%	–	–
Variation	28	1.40%	7	25%
Momentum	16	7.27%	–	–
Moving Average	20	10%	–	–
W. Moving Average	23	11.50%	–	–
RSI T.	40	0.96%	–	–
Total	6.057	2.268%	19	0.314%

Nevertheless, these agents do not prove clearly their ability to obtain this performance in exploiting some kind of inefficiencies. Said differently, once a *bootstrap reality check* procedure has been performed, we cannot provide evidence that their performance is not due to mere chance.

In this research, we focus on the simplest level of the implemented MAS, that is, *Family Agents* and *Strategic Agents*. This first step was necessary to investigate the weak-form EMH with a broad-spectrum design. Although we cannot provide here evidence of market inefficiencies, these results suggest that more complex agents, behaving like real-world technical traders, combining various indicators to shape their strategies, might obtain a very different result. This last part of our work, based on *Meta Agents*, has still to be perfected to capture, if at all possible, some anomalies in financial data. Nevertheless, this is a necessary next step if one wants to invalidate the usual objection coming from many chartists or technical traders when quantitative analysis refute their so-called ability to outperform the market: their "knowledge" is often presented as a combination of complex receipts, which make a scientific verification difficult. MAS and Artificial Intelligence may here be very useful to design strict and robust tests.

References

[1] Brock, W., Lakonishock, J., and LeBaron, B. (1992). Simple technical trading rules and the stochastic properties of stock returns. *Journal of Finance*.
[2] Detry, P. and Gregoire, P. (2001). Other evidences of the predictive power of technical analysis: the moving-average rules on european indices. In *Proceedings of the Lugano European Financial Management Association Conference*.

[3] Fama, E. (1970). Efficient capital markets: A review of theory and empirical work. *Journal of Finance.*

[4] Jensen, D. and Cohen, P. (2000). Multiple comparisons in induction algorithms. *Machine Learning.*

[5] Jensen, M. (1967). Random walks, reality or myth - comment. *Financial Analyst Journal.*

[6] Jensen, M. and Benington, G. (1969). Random walks and technical theories: Some additional evidence. *Journal of Finance.*

[7] Malkiel, B. (2003). The efficient market hypothesis and its critics. *Journal of Economic Perspectives.*

[8] Park, C. and Irwin, S. (2004). The profitability of technical analysis. *AgMAS project Research Report 2004-4.*

[9] Politis, D. and Romano, J. (1994). The stationary bootstrap. *Journal of the American Sta- tistical Association.*

[10] Samuelson, P. (1965). Proof that properly anticipated prices fluctuate randomly. *Industrial Management Review.*

[11] Shiller, R. J. (2003). From efficient markets theory to behavioral finance. *Journal of Economic Perspectives.*

[12] White, H. (2000). A reality check for data snooping. *Econometrica.*

[13] Wooldridge, M. (2002). *An Introduction to MultiAgent Systems.* John Wiley and sons.

5

The Dynamics of Quote Prices in an Artificial Financial Market with Learning Effects

Andrea Consiglio[1], Valerio Lacagnina[2], and Annalisa Russino[3]

[1] Dip. di Scienze Statistiche e Matematiche, Viale delle Scienze, Palermo, Italy.
   consiglio@unipa.it
[2] Dip. di Scienze Statistiche e Matematiche, Viale delle Scienze, Palermo, Italy.
   ricopa@unipa.it
[3] Dip. di Scienze Statistiche e Matematiche, Viale delle Scienze, Palermo, Italy.
   arussino@dssm.unipa.it

5.1 Introduction

In this paper we study the evolution of bid and ask prices in an electronic financial market populated by portfolio traders who optimally choose their allocation strategy on the basis of their views about market conditions.

Recently, a growing literature has investigated the consequences of learning about the returns process[4]. There has been an increasing interest in analyzing what are the implications of relaxing the assumption that agents hold correct expectations.

Under the assumption of rational expectations, agents know the true probability law underlying equilibrium economic variables, and, additionally, they know the functional relations between equilibrium magnitudes. Therefore, the market volatility depends uniquely on the variability of exogenous fundamental information about the external environment. This conclusion is in conflict with many empirical studies showing that financial markets exhibit typical anomalies like returns predictability, excess volatility with respect to the fundamentals, and volatility clustering.

Academic research has attempted to explain the market paradoxes introducing irrational behavior such as the existence of a certain number of noise traders (chartists), or investors who perceive probabilities incorrectly or are vulnerable to the impact of fads. In the last years, a new direction of research has recognized the importance of analyzing the effect of a learning process about the structural relationships of the economy. Learning introduces a link between state variables and agent's beliefs, and thus creates endogenous un-

---

[4] See Bossaerts [4], Barberis [1], Brennan and Xia [5], Xia [18], Kurtz and Motolese [14], Lewellen and Shanken [15], Guidolin and Timmermann [11]. See Hommes [12], for a recent survey on heterogeneous agent models.

certainty which depends on the distribution of agents' beliefs, on the frequency of change in agents' beliefs, and on the correlation among beliefs.

The majority of the literature on asset pricing under learning has been developed assuming a representative investor and (or) a single stock market.

By contrast, in this paper we model an artificial financial market whit multiple risky assets and heterogeneous beliefs. To analyze the process by which the dynamics of the learning process about the economy affect portfolio choices and market outcomes we set up a framework where the market volatility depends on the interaction between agents' beliefs and market results.

In particular, we design an order book market system where agents enter the market sequentially and trade to adjust their portfolio according to their optimal target allocations. We create heterogeneity assuming that investors have imperfect information about the joint distribution of returns. In particular, we allow agents to hold arbitrary priors about the univariate marginal distribution of returns, and we make agents update those distribution using past realized returns. We concentrate our attention on analyzing the impact of a learning process about the marginal distributions of returns assuming that agents have a constant common view of the assets' association structure. They correctly apply a copula function to generate the joint distribution of returns to be used to determine the optimal portfolio allocations. We assign to all agents the same investment horizon, but we create asynchronous updating assuming that different groups of agents entered the market at different moments in time. Finally, we simplify the optimization problem assuming that investors are myopic in the sense that, at the beginning of the investment horizon they choose their portfolios as if there will be no further trading. At the end of the investment horizon agents use the observed market prices to update the joint distribution of returns and choose their new optimal portfolio[5].

Automated systems offer advantages in terms of operating and trading costs, but they depend on public limit orders for the provision of liquidity. The time variation in liquidity can affect the evolution of prices, and a complex dynamics can arise between measures of market trading activity and measures of market volatility[6].

In this paper, we analyze how the dynamics of the distribution of beliefs over time affect market price changes. Moreover, we highlight the complex dynamics arising between time variation in trading activity, time variation in liquidity and quote price changes.

In Consiglio and Russino [7], using the same basic framework, we have shown that, under learning, the automated auction system generates irregular price series characterized by sharp increases and decreases (looking like bubbles and crashes), but that the jumps in the price series are not related

---

[5] See Consiglio and Russino [7] for the details of the model that we implement.

[6] The topic has been addressed by Domowitz in a series of papers analyzing the market behavior of real electronic markets [e.g., 9, 8, 10].

to sudden changes in the optimal portfolio weights. We analyzed the unconditional distribution of price changes and we provided evidence supporting the hypothesis that the parameters characterizing the learning process affect significantly the evolution of market liquidity, and that the variability of market liquidity determines the observed *bubble-like* phenomena. This paper is an extension of Consiglio and Russino [7] in two directions.

First, we analyze what is the role played by the assumed portfolio optimization model in affecting the market dynamics. That is, maintaining constant all the parameters governing the learning process, we compare two settings where we change the utility function assigned to the agents. In the first setting, we assume that investors' preferences can be represented by a standard mean-variance reduced utility function. In the second setting, we use a *prospect-type* [13] preferences. In particular, we model the utility function in terms of deviations, measured at regular intervals, from a specified target growth rate of wealth, and we assume that investors are more sensitive to downside movements. We run a series of simulations maintaining constant the parameters governing the learning process and we analyze the dynamics of market prices. We find that under mean-variance choices, for each asset, the optimal target allocations show lower variability across agents and over time, reflecting a lower sensitivity to the distribution of agents' beliefs and to its temporal variation.

Second, we study the time-series behavior of quote prices. We use intra–day hourly data relative to the order flow and the structure of the order–book to analyze the short-run impact of those variables on price changes. We estimate a VAR model for bid and price changes and we include, as exogenous regressors, variables measuring market trading activity and variables measuring market liquidity. The simultaneous analysis the time series behavior of quote prices allows to capture any asymmetries in the two series. We show that the dynamics of the best quotes is mainly affected by the variables measuring the market liquidity. The price series show self-regulating properties: extreme movements in prices induce contra–side order activity. Past changes in bid prices have a stronger effect on ask prices than viceversa, indicating that contra–side selling activity is particularly strong.

The paper is structured as follows. Section 5.2 describes the market model. Section 5.3 presents the calibration used for our simulations. Section 5.4 discusses the preliminary results obtained.

5.2 The Model

5.2.1 The Market Setting

We consider an economy with M agents and N risky assets. The market works as a double-auction automated system. Agents, trading to reach their optimal portfolio, enter the market sequentially. At each time step k within a trading day t we make the probability of entering the market for the i-th agent, $P_i(E)$, depend on the total imbalance between the target and the current portfolio,

$$P_i(E) = f(\Delta_i) \qquad (5.1)$$

$$\Delta_i = \sum_{j=1}^{N} \left| h_{ij}^*(t, t+\tau) - \frac{x_{ij}^t(k) \, P_j^t(k)}{W_i^t(k)} \right|$$

where $h_{ij}^*(t, t+\tau)$ is the agent's optimal target allocation for asset j, $x_{ij}^t(k)$ represents the agent's current holding in asset j, $P_j^t(k)$ is the current price for asset j, and $W_i^t(k)$ is the agent's total wealth given current prices and agent's holdings. Thus, the activation function $P_i(E)$ reflects the urgency of trading for the candidate agent. Agents are more impatient to trade, the more distant is their current wealth allocation from their target portfolio.

When a trader enters the market he faces an exchange book with orders to buy and to sell. Agents can trade immediately at the current quotes, placing *market orders*, or they can submit *limit orders* that are stored in the exchange book and will be executed if matching orders arrive before the end of the trading day. Limit orders will be executed using first price priority and then time precedence. At each moment in time during the day, the exchange book, divided in a buy side and a sell side, shows all the orders that have been issued up to that time and that have not found a matching order. For each order, the order size, the limit price, and the posting time are reported. Prices move in discrete steps, and, during each trading day, the minimum tick size depends on the daily opening price. At the end of the trading day all orders are cancelled. The spot price at each time step k is either the last transaction price or the last midquote, if a change in the quotes occurred.

5.3 The Agents' Behavior

Agents' behavior is specified in terms of order flow strategy (number of units to buy or to sell) and order–type submission strategy (market or limit order).

Agents trade to rebalance their portfolio. That is, at each moment in time they trade to adjust their portfolio according to their optimal target allocations. At time step k during trading day t, the number of units of the j-th asset that the i-th agent is willing to trade is given by,

$$q_{ij}^t(k) = \left\lfloor \frac{h_{ij}^*(t, t+\tau) \, W_i^t(k) - x_{ij}^t(k) \, P_j^t(k)}{P_j^t(k)} \right\rfloor \qquad (5.2)$$

where $\lfloor \cdot \rfloor$ denotes the integer part. If $q_{ij}^t(k) > 0$, the trader issues a buy order; if $q_{ij}^t(k) < 0$, the trader issues a sell order. The target allocations $h_{ij}^*(t, t+\tau)$, where τ represents the length of the investment horizon, are the optimal solutions of the agent's portfolio choice problem. Agents are cash constrained. In particular, borrowing and short-selling are not allowed and the agent's orders can be submitted only if the money needed (MN) is not greater than the money available (MA),

$$MN_i(k) = \sum_{j=1}^{N} I_{\{MB\}}^{j}(k)\, q_j'(k)\, A_j(k) + \sum_{j=1}^{N} I_{\{LB\}}^{j}(k) \left[q_{ij}(k) - q_j'(k)\right] P_{j,b}(k)$$

$$(5.3)$$

$$MA_i(k) = C_i(k) + \sum_{j=1}^{N} I_{\{MS\}}^{j}(k)\, q_j'(k)\, B_j(k)^7$$

$$(5.4)$$

where $I_{\{A\}}^{j}(k)$ is an indicator variable denoting for each risky asset j if the agent wants to issue an order; $q_j'(k)$ is the minimum between the quantity that the agents wants to trade at current prices, $q_{ij}(k)$, and the quantity available at the current quote, $Q_j(k)$; and $C_i(k)$ is the cash available to the i-th agent. The event A can be a market order to buy (MB), a limit order to buy (LB), or a market order to sell (MS). If $MN_i(k) \leq MA_i(k)$, then all the orders that the agent wants to issue will be submitted. If $MN_i(k) > MA_i(k)$, then for each asset j the number of units to trade is scaled down until $MN_i(k) = MA_i(k)$. The quantity adjustment keeps constant, with respect to the total, the percentage of money to allocate in each asset. The adjustments imposed by the budget constraint are performed giving priority to the submission of market orders to buy. Only if some money remains available after all market orders to buy have been processed, the procedure to check for the availability of money for submitting the desired limit orders to buy starts. Otherwise, the limit orders to buy are all cancelled.

We specify exogenously the order–type submission criterion. That is, we assume that traders want to satisfy their trading needs as soon as possible, and thus they will submit a market order at the current quote for the quantity they need to trade. Limit orders are used only if for some j the corresponding $q_{ij}^{t}(k)$ is greater than the quantity available at the current quote. In this case the agent places a market order for the quantity available, and for the residual quantity, given by $q_{ij}^{t}(k) - Q_j^{t}(k)$, he will submit a limit order. The associated limit price will be such that the order will be first on the appropriate side of the book, so we have that,

$$P_{j,b}^{t}(k) = B_j^{t}(k) + \epsilon^t$$
$$P_{j,s}^{t}(k) = A_j^{t}(k) - \epsilon^t$$

where $B_j^{t}(k)$ and $A_j^{t}(k)$ are respectively the best bid and the best ask in the order book, and ϵ^t is the minimum tick size for trading day t. When there are no orders on the relevant side of the book to match with, the agent will place directly a limit order for the whole quantity needed, $q_{ij}^{t}(k)$, at a price that will make him first on the book.

---

7 To simplify the notation we drop the superscript indicating the trading day

68   Andrea Consiglio et al.

5.3.1 The Learning Process

We assume that investors have imperfect information about the joint distribution of returns, and that they must learn about the unknown returns
generating process using the available information. In particular, we allow
agents to hold arbitrary marginal prior densities for the assets returns. We
model the prior marginal returns distribution of each asset as a Dirichlet with
parameters $(\alpha_1, \ldots, \alpha_C)$ where C is number of classes of the support of the
returns distribution. Thus, we assign to the agents populating our economy
arbitrary prior densities given by,

$$f_{ij}(\theta) = \frac{\Gamma(\alpha_{ij1} + \ldots + \alpha_{ijC})}{\Gamma(\alpha_{ij1}) \ldots \Gamma(\alpha_{ijC})} \theta_1^{\alpha_{ij1}-1} \ldots \theta_C^{\alpha_{ijC}-1} \qquad (5.5)$$

where $\theta_1, \ldots, \theta_C \geq 0$; $\sum_{c=1}^{C} \theta_c = 1$, $i = 1, \ldots, M$, and $j = 1, \ldots, N$. Agents
will use the history of observed market returns to update their beliefs in a
bayesian fashion. Letting v_jc be the number of returns observed, for the j-
th asset, in class c during the time period between two successive updating
days, the posterior distribution of the i-th agent for the returns of asset j will
be Dirichlet with parameters $((\alpha_{ij1} + v_{ij1}), \ldots, (\alpha_{ijC} + v_{ijC}))$. To determine
the optimal portfolio composition, agents should know the joint probability
distribution of assets returns. We assume that agents share a common constant view of the securities association structure, and that they correctly use
a copula function[8] to generate the N-variate returns distribution from their
arbitrary set of N univariate distributions. We use The Gaussian copula to
model the dependence structure between the risky asset. Given their own
marginal univariate returns distributions and the assigned copula, agents extract from the multivariate distribution of returns a number S of scenarios.
Each scenario specifies a return for each of the N risky assets for all the time
periods in the investment horizon of the agent. Every scenario represents a
possible future realization of returns, for the N assets, given the agent's joint
probability distribution. Agents use the S extracted scenarios to determine
the optimal composition of their portfolio.

5.3.2 The Portfolio Model

We compare two cases. In the first case, we assume that investors, consistently with their rational learning process, choose their portfolio maximizing
their expected utility of the end of horizon financial wealth. We assign to the
agents a standard mean–variance expected utility function so that the agent's
portfolio problem is simply,

$$\max_{\mathbf{h}} E[U(\tilde{W}_T)] = \max_{\mathbf{h}} \{\mu'\mathbf{h} - \frac{\lambda}{2}\mathbf{h}'\Sigma\mathbf{h} \mid \mathbf{i}'\mathbf{h} = 1; \mathbf{h} \geq 0\} \qquad (5.6)$$

---
[8] See [17, 16].

where μ and $\mathbf{h}$ are the n-vectors of rates of returns and target allocations, Σ is an (n, n)-positive definite covariance matrix, and λ is the risk aversion parameter.

In the second case, we introduce an element of irrationality assigning to the agents prospect–type preferences [13]. We assume that each investor has an initial level of wealth and a target growth rate to reach within his investment horizon. The investor must determine an asset allocation strategy so that the portfolio growth rate will be sufficient to reach the target. We model the utility function in terms of deviations, measured at regular intervals, from the specified target, and we assume that investors are more sensitive to downside movements. Our approach is inspired to the descriptive models about investors' choices followed by [3, 2]. The target portfolio holdings are determined using the scenario optimization model developed in [6].

Let $\tilde{u}_t$ and $\tilde{d}_t$ be two random variables which define the upsides and downsides in each time period $t = 1, 2, \ldots, T$, the random variables $\tilde{D}_T$ and $\tilde{U}_T$ which accounts respectively for the final deficit and surplus are given by,

$$\tilde{D}_T = \sum_{t=1}^{T} \tilde{d}_t$$
$$\tilde{U}_T = \sum_{t=1}^{T} \tilde{u}_t$$

The investor will determine his portfolio by solving the following multi-objective programming model

$$\text{Maximize}_{\mathbf{h}} \qquad E[\tilde{U}_T] - \lambda \, E[\tilde{D}_T] \qquad (5.7)$$
$$\text{s.t.}$$
$$\mathbf{h} \in \mathcal{H}, \qquad (5.8)$$

where $\lambda > 0$ is the loss aversion parameter.

5.4 Results

5.4.1 Simulation Parameters

We define a setting where agents are equal in terms of endowments, trading strategies, investment horizons, risk profiles, institutional constraints, and type of information used to update prior beliefs. The main source of heterogeneity is given by the different set of prior univariate marginal distributions of returns that we assign to each group of agents.

We run our simulations with a population of $M = 6000$ potentially active traders, $T = 2280$ trading days, and $N = 3$ risky assets. Each trading day is divided in $K = 360$ time steps corresponding to a trading day of six hours, assuming a time step k equal to one minute. Every agent gets an initial endowment in each of the N stocks of our economy of 50 shares, and a cash endowment of $C_i = \text{EUR}1000$. Initial prices are set equal to $100 \, €$. Agents are divided in $G = 6$ equally sized groups. All the agents in a group share the same view about the joint distribution of returns. We create agents' heterogeneity assigning to each group of agents a different set of prior univariate marginal distributions of returns. We divide the population in pessimists and optimists. That is, we assume that the univariate marginal priors reflect agents expectations about market performance. Thus, a group of pessimistic agents will have a set of prior distributions, one for each of the N risky assets, with a support shifted to the left, while a group of optimists will have priors with support shifted to the right. The constant correlation matrix used by all agents to obtain the joint distribution of returns can be of two types: one with negative correlations among the three assets, and one with positive correlations.

All agents have the same type of objective function, the same risk measure $\lambda = 2.25$, and the same investment horizon of $H = 240$ days, corresponding to one year assuming a trading week of five days. The target portfolio return, g_i, is different between pessimistic and optimistic agents: $g_i = 10\%$ for pessimists and $g_i = 20\%$ for agents with optimistic views. To maintain an active market over time, we simulate a setting where agents have not entered the market at the same time. That is, we assume that every $I = 20$ days, one group reaches the end of his investment horizon, and all the agents in the group update their view about the joint distribution of returns using the history of observed returns. The posterior distribution of returns is then used to determine new optimal target allocations $h_{ij}^*(t, t+\tau)$. To create a history of prices that agents can use to update their priors, for the first 240 days, we run our simulations, using randomly assigned target allocation vectors. That is, during the initial period each group of agents gets target allocation vectors sampled from a Dirichlet($1,\ldots 1;1$). To maintain, over the entire length of the simulations, a market structure homogeneous in terms of trading activity, during the initial period, we randomly extract one group of agents every $I = 20$ days, and we assign to all the agents in the selected group new random target allocations. The minimum tick size is 1% of the opening daily price.

5.4.2 Comparison Between the Portfolio Models

To compare the effect of learning on prices under the two different portfolio models, we run simulations where we maintain constant all the parameters and the seed number but we change the portfolio model used by the agents.

Fig. 5.1 shows the times series of the prices of the three assets generated in the two different settings. In the upper panel we plot the price series corresponding to the case where agents use a reduced mean–variance utility function (MV) both for the case of negative association structure among the N risky (left, NMV) and the case of positive association (right, PMV). In the botttom panel we display the time series of prices determined under the prospect–type preferences (TR) in the two cases of negative and positive association (left, NTR and right panel, PTR).

Clearly, the price series generated under prospect-type preferences are more irregular, and present sharp upward and downward movements. The dif-

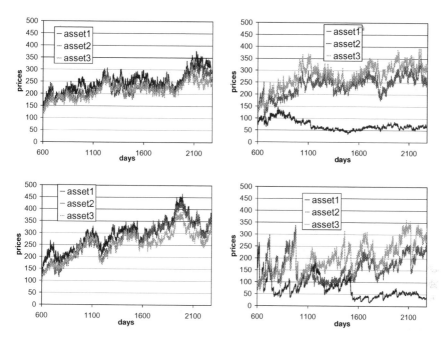

Fig. 5.1. Daily time series of prices for the different settings. From the upper–left to the right–bottom panel, we display the NMV, PMV, NTR, and PTR

ference becomes more pronounced in the series generated assuming a positive association structure among the risky assets. The positive correlation setting is the setting where the agents have less scope for diversification and, thus, where agents' choices are more influenced by the marginal univariate distribution of returns. The visual inspection of the plots suggests that prospect-type preferences make agents' choices, and consequently prices, much more reactive to agents' views. We build an aggregate measure of the agents' heterogeneity in terms of optimal allocation, every $I = 20$ days, summing the euclidian distance between each possible pair of target allocation vectors. Since there are 12 groups, we take the sum of 66 euclidian distances. Similarly, we build a measure of the agents' heterogeneity in terms of views summing, every $I = 20$ days, the Δ-Distance between each possible pair of joint distributions of returns. Given two joint distributions discretized in N classes and letting p_i and q_i be respectively the probability of having one observation in the i-th class under the two joint distributions the Δ-Distance is given by,

$$\Delta(p,q) = \sum_{i=1}^{C} \frac{(p_i - q_i)^2}{p_i + q_i}$$

In Fig. 5.2 we plot the time series of the aggregate euclidian distance between optimal allocation vectors and of the aggregate Δ-Distance between the joint

distributions of returns for the two cases of mean-variance and prospect preferences. Clearly, the portfolio model based on prospect preferences makes the
action choices extremely sensitive to the agents' views. Agents optimal allocation vectors are more variable across agents over time reflecting the agents
diversity in terms of beliefs. The greater heterogeneity of agents' optimal target allocation induced in the market setting based on prospect preferences
generates a more intense trading activity and that in turn produces irregular
price series.

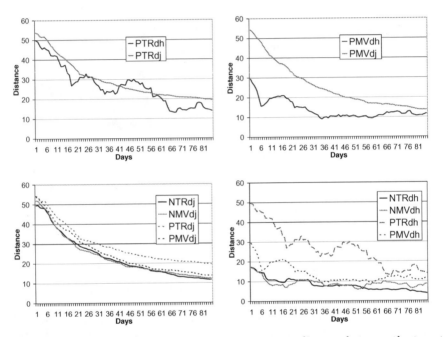

Fig. 5.2. In the upper panels we report the aggregate distance between the target
allocations and the joint distributions across groups over time for the PTR setting on
the left, and for the PMV setting on the right. The bottom panels show the temporal
behavior of the aggregate distance between the joint distributions (left), and of the
aggregate distance between the target allocations (right) for all the settings.

5.4.3 Quote Price Dynamics

In an automated market system the temporal behavior of prices depends on
the endogenous ability of the market to maintain liquidity, generically defined
as the market ability to closely pair the desires of buyers and sellers. The
temporary lack of liquidity and accumulation of orders on one side of the book
can generate jumps in prices. To highlight the relations between price changes,

trading activity, and liquidity, we analyze the dynamics of ask and bid quotes in the setting with prospect-type preferences and positive association structure among the risky assets (PTR). As we have seen, this setting is characterized by the highest heterogeneity across agents over time, and therefore is the setting generating the highest trading activity. Using hourly data for 480 days, we estimate a VAR model for logarithmic change of the best bid and of the best ask prices. We specify as exogenous variables, a set of variables measuring the liquidity of the market (the log-spread, S_t, and the imbalance between the queues of the buy and the sell side of the book, ΔQ_t), and a set of variables measuring the trading activity (the number of transactions, M_t, the trading volume, V_t, and the average waiting time between transactions, WT_t). We interpret missing values of the best ask and of best bid series as cases where the market price is so low or so high (in the case of the bid price) that nobody is willing to trade. Thus, we substitute missing values of the two series respectively with the minimum and the maximum value of the series. As we can see from Table 5.1 the two series are highly and persistently negatively auto–correlated. Interestingly, the two series are also negatively cross–correlated: an increase in the best bid will induce after same time a decrease in the best ask and viceversa. The impact of changes in the best quote of the opposite side of the market is stronger in the case of ask prices. All that suggest that the market displays self-regulating properties: waves of agents on one side of the market causing extreme movements in prices are followed by contra–side order flow activity. The phenomenon is stronger in the case of up movements in prices. Generally, the evolution of the best quotes is affected mostly by the variables related to market liquidity. The log-spread has a contemporaneous negative effect on ask prices and a positive contemporaneous effect on bid prices. Additionally, increases of the log-spread lead after some time to increases in the ask quote and to reductions in the bid quote. All that suggest that increases in the spread are related to the occurrence of either selling waves depleting the bid book, or buying waves that scale up the sell side of the order book. Lastly, past imbalances between the buy and the sell queue of the order book have a significant persistent positive effect on both ask and bid prices. The variable measuring the trading activity are less important to explain the evolution of ask and bid prices. The fact that the trading volume is negatively related to changes in ask prices and that the number transactions is negatively related to changes in bid prices suggest that an increase in trading activity has an influence on quote prices when it is caused by selling transactions.

Table 5.1. Results of the VAR model estimated for the log–changes in quote prices in the PTR setting. On the left we report the results for the log–changes of the best ask. On the right we report the results relative to log–changes of the best bid.

		Coef.	Std. Err.	z	P > |z|			Coef.	Std. Err.	z	P > |z|
ΔA	ΔA					ΔB	ΔA				
	t−1	-.63821	.01913	-33.36	0.000		t−1	-.03478	.01570	-2.22	0.027
	t−2	-.46144	.02075	-22.24	0.000		t−2	-.05118	.01703	-3.01	0.003
	t−3	-.29403	.02055	-14.31	0.000		t−3	-.01967	.01686	-1.17	0.243
	t−4	-.20264	.01857	-10.91	0.000		t−4	-.00930	.01524	-0.61	0.542
	t−5	-.10818	.01514	-7.14	0.000		t−5	.00354	.01243	0.29	0.776
	ΔB						ΔB				
	t−1	-.05920	.02354	-2.51	0.012		t−1	-.65867	.01932	-34.10	0.000
	t−2	-.07276	.02548	-2.86	0.004		t−2	-.44952	.02091	-21.50	0.000
	t−3	-.00864	.02524	-0.34	0.732		t−3	-.30637	.02071	-14.80	0.000
	t−4	.02766	.02337	1.18	0.237		t−4	-.21214	.01918	-11.06	0.000
	t−5	-.00171	.01909	-0.09	0.928		t−5	-.11197	.01566	-7.15	0.000
	S						S				
	t	-6.15645	.22866	-26.92	0.000		t	5.14180	.18762	27.41	0.000
	t−1	1.91701	.29868	6.42	0.000		t−1	-2.08803	.24508	-8.52	0.000
	t−2	1.13451	.29154	3.89	0.000		t−2	-1.31587	.23922	-5.50	0.000
	ΔQ						ΔQ				
	t	-.02870	.00523	-5.49	0.000		t	.00156	.00429	0.36	0.715
	t−1	.02482	.00595	4.17	0.000		t−1	.02058	.00488	4.22	0.000
	t−2	.04457	.00539	8.26	0.000		t−2	.03995	.00443	9.03	0.000
	M						M				
	t	.00214	.00133	1.60	0.109		t	-.00446	.00109	-4.08	0.000
	t−1	-.00056	.00133	-0.42	0.677		t−1	.00012	.00109	0.11	0.915
	V						V				
	t	-.00027	.00010	-2.74	0.006		t	-.00007	.00008	-0.84	0.403
	t−1	.00011	.0001	1.11	0.266		t−1	.00015	.00008	1.81	0.071
	WT						WT				
	t	-.00076	.00012	-6.42	0.000		t	-.00018	.00010	-1.84	0.066
	t−1	.00091	.00012	7.69	0.000		t−1	-.00033	.00010	-3.42	0.001
	c	.07997	.01769	4.52	0.000		c	.00103	.01452	0.07	0.943

References

[1] Barberis, N. (2000). Investing for the Long Run when Returns are Predictable. *Journal of Finance*, 55:225–264.

[2] Barberis, N., Huang, M., and Santos, T. (2001). Prospect Theory and Asset Prices. *Quarterly Journal of Economics*, CXVI:1–52.

[3] Benartzi, S. and Thaler, R. (1995). Myopic Loss Aversion and the Equity Premium Puzzle. *Quarterly Journal of Economics*, 110(1):73–92.

[4] Bossaerts, P. (1999). Learning–Induced Securities Price Volatility. Working Paper, California Institute of Technology.

[5] Brennan, M. and Xia, Y. (2001). Stock Price Volatility and Equity Premium. *Journal of Monetary Economics*, 47:249–283.

[6] Consiglio, A., Cocco, F., and Zenios, S. (2004). www.Personal_Asset_Allocation. *Interfaces*, 34:287–302.

[7] Consiglio, A. and Russino, A. (2006). How Does Learning Affect Market Liquidity? A Simulation Analysis of a Double-Auction Financial Market

with Portfolio Traders. *Journal of Economic Dynamics and Control.* forthcoming.

[8] Coppejans, M., Domowitz, I., and Madhavan, A. (2001). Liquidity in an Automated Auction. Working Paper.

[9] Domowitz, I. and El-Gamal, M. (1999b). Financial Market Liquidity and the Distribution of Prices. Working Paper 179950, ssrn.

[10] Domowitz, I. and Wang, X. (2002). Liquidity, Liquidity Commonality and Its Impact on Portfolio Theory. Working Paper.

[11] Guidolin, M. and Timmermann, A. (2005). Properties of Asset Prices Under Alternative Learning Schemes. Working Paper 009, Federal Reserve Bank of St. Louis.

[12] Hommes, C. H. (2005). Heterogeneous Agent Models in Economics and Finance. In Judd, K. L. and Tesfatsion, L., editors, *Handbook of Computational Economics*, volume 2. Elsevier Science.

[13] Kahneman, K. and Tversky, A. (1979). Prospec Theory: an Analisys of Decision under Risk. *Econometrica*, 47(2):263–291.

[14] Kurtz, M. and Motolese, M. (2001). Endogenous Uncertainty and Market Volatility. *Economic Theory*, 17:497–544.

[15] Lewellen, J. and Shanken, J. (2002). Learning, Asset-Pricing Tests, and Market Efficiency. *The Journal of Finance*, LVII:1113–1145.

[16] Malevergne, Y. and Sornette, D. (2003). Testing the Gaussian Copula Hypothesis for Financial Assets Dependencies. *Quantitative Finance*, 3:231–250.

[17] Nelsen, R. (1999). *An Introduction to Copulas*. Springer-Verlag, New York.

[18] Xia, Y. (2001). Learning about Predictability: The Effect of Parameter Uncertainty on Dynamic Asset Allocation. *The Journal of Finance*, 56:205–246.

6

Reduction of the Bullwhip Effect in Supply Chains Through Speculation

Thierry Moyaux and Peter McBurney

Department of Computer Science, University of Liverpool, Liverpool L69 3BX, UK, {moyaux,peter}@csc.liv.ac.uk

Summary. Agent-based simulations show that some kinds of speculators are able to stabilize the price in a market and to make this market more efficient. Instead of a single market, we consider a supply chain comprising a sequence of three markets in order to check that such speculators can also stabilize a supply chain. Specifically, we verify if these speculators reduce the price fluctuations caused by a phenomenon called the bullwhip effect, which is the amplification of order variability in supply chains. Our simulations show that speculation reduces such price fluctuations, even if price bubbles may appear. Another point is that the speculators we consider lose money in reducing these fluctuations while all the other agents would get richer and richer when the equilibrium is achieved in every market of the supply chain.

6.1 Introduction

Empirical evidence shows that orders in a supply chain are more variable for raw material producers than the initial demand addressed by end-customers to retailers. This phenomenon of amplification of order variability is known as the bullwhip effect [3]. The problem with the bullwhip effect is not only its essence itself, i.e., demand becomes more variable along the supply chain, but also the fact that it makes demand less predictable. Both increased variability and unpredictability cause important financial costs due to higher inventory levels and agility reduction. As an insight into the importance of these inefficiencies, Carlsson and Fullér [1] claim that the bullwhip effect would cost 100-200 MFIM (17-34 million euros) per year to the Finnish forest products industry, which has a total turnover of more than 100 BFIM (17 billion euros). The solution most often proposed to the bullwhip effect is information sharing [9], but other proposed solutions have included: EDLP (Every Day Low Pricing) policy or the allocation of sales based on past sales [3]. See [6] for a literature review of known causes with their solutions. In this paper, we study whether the presence of speculators could serve as another solution, which seems to be an approach never investigated before. More generally, we investigate interactions between related markets.

To this end, we see our supply chain as a sequence of markets, then we apply approaches stabilizing a single market to stabilize our supply chain. Steiglitz and his colleagues [5, 10, 11] provide us with such market-stabilizing approaches. In fact, they analysed the behaviour of a quite general market in which they noted that the presence of speculators (agents who simply try to buy low and sell high) are beneficial to all agents[1]. Specifically, adding the two considered kinds of speculators stabilizes the price of the only considered good[2], and this stabilization *"results in an overall increase in market efficiency and fluidity, in the sense that individual production decisions are more closely matched to skill, and the numeraire is more easily converted into accumulated wealth"* [10, p. 3]. In this paper, we study whether one of these same two kinds of speculators also stabilizes supply chains, in order to have a new solution to the bullwhip effect. Our concern deals with neither the reason for speculation (e.g. is speculation either an irrational behaviour, a rational behaviour due to asymmetric information, or a rational behaviour due to different degrees of risk aversion [4]) nor its potentially harmful consequences (e.g., bubbles, crashes and continued high trading volume [7]), but we only focus on the potential benefits of the aforementioned type of speculators for a supply chain.

This paper is organized as follows. Section 6.2 presents the simulation model. Then, Section 6.3 outlines the results obtained with this model. Section 6.4 presents how our speculators behave in a reproduction of their original environment proposed in [10, 11]. Section 6.5 discusses our approach.

6.2 Simulation Model

6.2.1 Overview of the Model

We have implemented our supply chain as three inter-linked marketplaces along which a single type of product is traded. As shown in Figure 6.1, these marketplaces are linked by companies, that is, paper mills and sawmills buy in one market and sell in another.

To be precise we use the settings outlined in Figure 6.1, that is, 25 end-customers buy furniture sold by 6 paper mills, which buy lumbers from 4 sawmills, which buy raw wood from 2 raw material suppliers. Figure 6.1 also shows that speculators buy and sell in the same market. We use the open-source JAVA Auction Simulator (JASA) [8] to implement each of the three marketplaces. Specifically, JASA provides several types of auctions (ascending auction, double auction, etc.) to represent each of our three markets, and, for the moment, we use a double auction for every marketplace. We have

---

[1] Kaldor [2] presents a defence of speculation in terms of economic stability.

[2] In this model, agents exchange gold for food, consumes one unit of food per day, and produces gold and food. Production skills of gold and food are different from each other, specific to individuals and constant over a simulation.

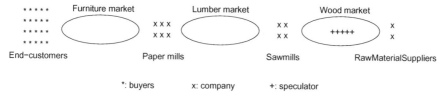

Fig. 6.1. Supply chain structure

extended JASA to have our three inter-linked marketplaces so that, at every round, the furniture market is activated first (i.e., end-customers may place a bid shout while paper mills may place an ask shout, then the furniture auctioneer calculates the clearing price, and finally physical exchanges take place at that price), next the lumber market, then the wood market, and finally a new round starts with the furniture market.

Let us now outline the two types of agents in our model, namely, companies and speculators. As can be seen in Figure 6.2, the representation of companies complies with the first level of the model SCOR from the Supply Chain Council [12], that is, companies are made of three functions:

(i) deliver is implemented as an inventory of finished products and a function to place ask shouts to sell products, (ii) make is the work-in-process inventory in which batches of items have to spend the production time before moving from the raw material inventory to the finished product inventory, and (iii) source is similar to deliver, except that its inventory contains raw materials and its function places bid shouts to buy products. Both asking and bidding functions depend on the capacity of their respective inventory. The asking function is the same as the bidding function to which we add a constant called *margin* because products have to be sold at a higher price than what they were bought plus production cost. Basically, a company is

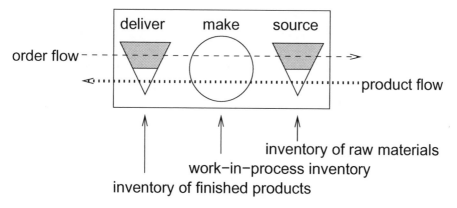

Fig. 6.2. Model of company.

thus represented by at least eleven parameters:

- 3 inventory capacities, i.e., one per inventory (but more when companies buy, sell and produce different types of products, which is not the case in this paper);
- 1 production rate (but more in general, when companies produce different types of products);
- 1 production cost called *margin* (but more in general, when companies produce different types of products);
- 3 parameters of the bidding function, because we use an adaptation of Steiglitz and his colleagues' function [5, 10, 11] which has three parameters;
- 3 parameters of the asking strategy, for the same reason as for the bidding strategy.

For simplicity, transportation is supposed instantaneous, production rate and inventory capacities are the same among agents, and, for every agent, the bidding function uses the same parameters as the asking function. More precisely, *an agent is charaterized by the three parameters* of its bidding function (the asking function having the same parameters).

6.2.2 Details of the Different Types of Companies

Whenever possible, we use the same parameters as presented by Steiglitz and Sha-piro [11], e.g. the initial funds of all agents is 60 units, and the initial (source or deliver) inventory level is uniformly randomly distributed between 15 and 20 units. The initialisation of every parameter is the same for all the companies in all the results presented in this paper. In other words, we do not change the seed of the pseudo-random number generators, so that Agent 0 (an end-customer) always begins with 17 units in its source inventory while Agent 1 (another end-customer) always stars with 19 units. Because of our adaptation of Steiglitz et al.'s model [5, 10, 11] to supply chains, additional parameters are necessary. The main ones are as follows:

- *End-customers* consume 1 unit of furniture per week, which is similar to [11], except that only end-customers are consumers. In order to afford this furniture, end-customers produce 5 units of money per week. Conversely to Steiglitz et al.'s model [5, 10, 11], only end-customers produce money, and they produce that money every week without having to choose what to produce.
- *Manufacturers* (i.e. paper mills and sawmills) have a capacited make function. Every manufacturer can produce a batch of exactly 10 units every week. They produce a new batch of products whenever possible, that is, every week in which 10 items are available in their source inventory.
  Manufacturers have a parameter *margin* in order to sell at a higher price than what they previously paid for the items to process. More precisely, when they place a ask shout, they add *margin* to the price that Steiglitz

et al.'s model [5, 10, 11] would normally choose[3]. Currently, sawmills have *margin* = 2 and paper mills *margin* = 3. We will see that this *margin* moves up the equilibrium price in every market of the supply chain.

- *Raw material suppliers* are special manufacturers because they also produce 10 units every week, except that they do not need to buy components first. In fact, raw material suppliers are the opposite from end-customers, that is, every end-customer is an infinite source of money and a product sink, while each raw material supplier is an infinite source of products and a money vacuum cleaner (but not a money sink).

  As paper mills and sawmills, raw material suppliers have a *margin* set to 1.

6.2.3 Details of the Speculators

Conversely to companies, speculators buy and sell in the same market. We reimplemented the same two types of speculators as Steiglitz *et al.* [10, 11]. Like other agents, speculators have no inventory holding cost, and they do not look for selling items as soon as possible. All speculators own 60 units of money at the beginning, which is the same amount as the companies. We use the type of speculators called AVG (for average) which use a moving average to update their forecast of the price of the product. If we call P the actual price and $\hat{P}$ the forecasted price, then the forecast of the current price is $\hat{P}(t) = \beta.\hat{P}(t-1) + (1-\beta).P(t-1)$ for some coefficient $\beta \in (0,1)$. Then, a ask shout $P(t-1).(1+margin)$ for all the owned products is placed when $P(t-1) < \hat{P}.(1-margin)$, and a bid shout $P(t-1).(1-margin)$ for all affordables products when $P(t-1) > \hat{P}.(1+margin)$. As in [11], $\beta = 0.008$ and the *margin* of the speculators is fixed at the beginning of a simulation by uniformly dividing the interval $[0.0, 0.5]$.

6.3 Results

6.3.1 Reproduction of the Bullwhip Effect

Before studying how speculators may reduce the bullwhip effect, we first present how this phenomenon shows in our model. Figure 6.3 displays the price at which good are exchanged every week (in every subfigure, time is displayed between Week 0 and 500, and price between 0 and 80). Specifically, you can see in Figure 6.3(a) that the price of wood is more stable than the price of lumber in Figure 6.3(b) which is itself less variable than the price of wood in Figure 6.3(c). These three subfigures do not reflect the bullwhip

---

[3] This is a first step. In future work, the sell price of finished products will depend on the buy price of its components, or at least, on the current price of these components.

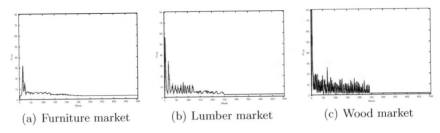

(a) Furniture market   (b) Lumber market   (c) Wood market

Fig. 6.3. Amplification of price fluctuations in the supply chain without speculators (the three subfigures have the same scale)

effect itself but its consequence on price. Since the bullwhip effect is defined as the amplification of order variability, we should use the quantity requested in bid shouts to describe this effect. However, since all agents have the same form of utility function, we think the shape of the figures would be the same.

As previously said in Footnote 3, the price in one market has no direct impact on the price in another market. That is, a company may buy expensive components in order to produce then sell cheap products. We will implement in future work such a "price stream" in our model. At the moment, companies are only linked through a stream of orders/bids and a stream of products. However, Figure 6.3 shows that the interactions between these two streams is enough to create a bullwhip effect which causes greater price fluctuations in wood market than in furniture market.

6.3.2 Impact of Speculators on the Price Fluctuations Caused by the Bullwhip Effect

Next, we add speculators in the lumber and wood markets. We do not add speculators in the furniture market because of the characteristics a good needs to possess to be speculated. In fact, according to Kaldor [2, p. 20], these "attributes are: (1) The good must be fully standardised, or capable of full standardisation; (2) It must be an article of general demand; (3) It must be durable; (4) It must be valuable in proportion to bulk." As a consequence, raw materials, such as the lumbers and wood, are more likely to be speculated, conversely to finished products as furniture.

Figure 6.4 presents price fluctuations in every market when 5 speculators trade in the lumber and/or the wood markets (specifically, there are 5 wood speculators in the supply chain outcomes presented in Subfigures 6.4(a), 6.4(b) and 6.4(c), 5 lumber speculators in Subfigures 6.4(d), 6.4(e) and 6.5(f), and 5 lumber speculators and 5 wood speculators in Subfigures 6.4(g), 6.4(h) and 6.4(i)), and, similarly, Figure 6.5 shows price evolution when 25 speculators trade in the lumber and/or wood markets.

We can note that the price in the wood market stabilizes when wood speculators are added, because the price in Subfigures 6.4(c) and 6.5(c) is

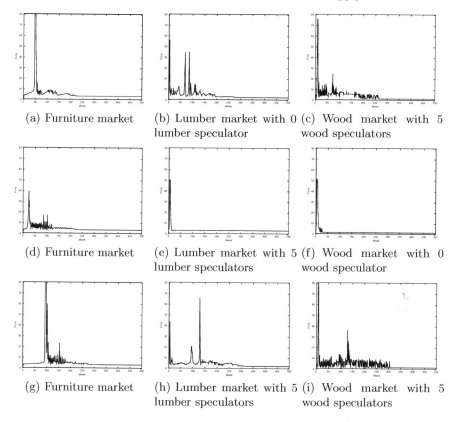

(a) Furniture market

(b) Lumber market with 0 lumber speculator

(c) Wood market with 5 wood speculators

(d) Furniture market

(e) Lumber market with 5 lumber speculators

(f) Wood market with 0 wood speculator

(g) Furniture market

(h) Lumber market with 5 lumber speculators

(i) Wood market with 5 wood speculators

Fig. 6.4. Amplification of price fluctuations in the supply chain with 5 speculators (all subfigures have the same scale as Figure 6.3)

more stable than in Subfigure 6.3(c). Moreover, 25 speculators have a greater stabilising effect than only 5, because the price in Subfigures 6.5(c) is more stable than in Subfigures 6.4(c). When we compare these same two subfigures, we can also notice that 25 speculators allow the price to stabilise quicker than with only 5 speculators. The same conclusion may be drawn when speculators are only added to the lumber market (cf. Subfigures 6.3(b), 6.4(e) and 6.5(e)). Since this paper addresses the impact of speculators on the bullwhip effect, we can first conclude that the type of speculators we consider is able to *stabilise the price fluctuations caused by the bullwhip effect.*

However, the smaller and shorter price fluctuations brought by speculators seem to be (sometimes) replaced by price bubbles of large amplitude. For instance, Subfigure 6.5(c) is much more stable than Subfigures 6.3(c), except that it has a single price bubble which is huge. Indeed, while no speculators lead to price fluctuations, too many speculators lead to price bubbles: there

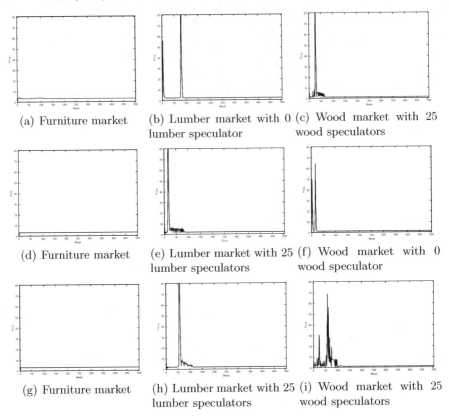

(a) Furniture market

(b) Lumber market with 0 lumber speculator

(c) Wood market with 25 wood speculators

(d) Furniture market

(e) Lumber market with 25 lumber speculators

(f) Wood market with 0 wood speculator

(g) Furniture market

(h) Lumber market with 25 lumber speculators

(i) Wood market with 25 wood speculators

Fig. 6.5. Amplification of price fluctuations in the supply chain with 25 speculators (all subfigures have the same scale as Figure 6.3)

seems to be an optimal number of speculators for every market in the supply chain. The question is thus how many speculators to use and where. For example, Subfigures 6.4(d), 6.4(e) and 6.4(f) are the most stable among all the subfigures in Figures 6.3, 6.4 and 6.5, while they do not use the largest number of speculators. On the contrary, the graphs in Figure 6.5 are more stable than in Figure 6.4 (i.e., if you compare Subfigure 6.4(n) with Subfigure 6.5(n)), except that there are a few price bubbles of large amplitude.

On the other hand, adding 5 wood speculators to Subfigure 6.4(f) destabilizes the wood market in Subfigure 6.4(i), which next destabilizes the lumber market in Subfigure 6.4(h). This seems to confirm that our simulation model allows to make a *link among the stability of different markets*. That is, speculators first stabilize their market, then other markets. And also, speculators first create a price bubble in their market, which next destabilises another market.

Supply chain level	0 lumber speculator & 0 wood speculator	
	μ	σ
End-customers	1,112.4	627.4
Paper mills	285.2	768.3
Sawmills	640.8	1,198.0
Raw material suppliers	18,793.0	592.2
Lumber speculators		
Wood speculators		

Supply chain level	0 lumber speculator & 5 wood speculators		5 lumber speculators & 0 wood speculator		5 lumber speculators & 5 wood speculators	
	μ	σ	μ	σ	μ	σ
End-customers	1,152.4	622.2	956.5	602.3	960.8	689.6
Paper mills	321.6	705.8	1,915.2	3,117.2	560.6	1,138.4
Sawmills	962.2	1,113.8	3,661.3	4,260.2	923.0	1,098.1
Raw material suppliers	17,702.3	38.8	9,718.6	33.1	19,206.0	421.4
Lumber speculators			96.6	49.0	0.2	3.7
Wood speculators	- 4.2	7.5			156.4	237.9

Supply chain level	0 lumber speculator & 25 wood speculators		25 lumber speculators & 0 wood speculator		25 lumber speculators & 25 wood speculators	
	μ	σ	μ	σ	μ	σ
End-customers	2,072.1	593.2	1,892.5	640.6	2,028.1	497.0
Paper mills	- 137.6	248.3	471.2	1,241.5	- 31.4	92.7
Sawmills	1,457.3	3,130.2	970.4	1,281.5	560.7	1,131.5
Raw material suppliers	7,273.0	86.6	8,626.9	366.7	9,434.1	277.1
Lumber speculators			- 4.2	8.3	- 3.7	9.4
Wood speculators	- 7.3	10.7			45.5	102.4

Fig. 6.6. Average funds and standard-deviation of funds between parentheses per level in the supply chain at Week 500.

Final, every market of the supply chain always reach its equilibrium price within 500 weeks. This equilibrium price is defined by the *margin* of the company selling in the considered market. As a consequence, the equilibrium price is 1 in the wood market, 2 in the lumber market and 3 in the furniture market.

6.3.3 Financial Aspects of Speculation in our Supply Chain Model

Another important point is the financial impact of speculation on companies and the cost efficiency of speculators. The table in Figure 6.6 presents the average amount and the standard-deviation of the money owned by the companies in every level of the supply chain. Three points can be noticed in this table:

- *The money ends up with the raw material suppliers:* The end-customers produce money which is next transferred to the raw material producers through the paper mills and the sawmills (and possibly via the speculators). These latter companies do not retain much of the money which goes through them. Of course, if simulations were run for a longer duration, the paper mills and sawmills would earn more and more money since, at the equilibrium of every market, they sell at one unit of money more expensive than they buy. This benefit is allowed by the use of different *margin*, which, as we have already said, move up the equilibrium price. As a consequence, many paper mills and sawmills have a negative amount of money in Week 500, but they would earn an infinite amount of money if the simulation was run forever. Nevertheless, they would still remain poorer than the raw material producers

- *The end-customers seem to produce money at a good rate:* In the considered first 500 weeks, producing 5 units of money per week seems to be the good rate because all end-customers have positive funds in Week 500. Since the equilibrium price of the furniture market is only 3 (because of the *margin* of paper mills), end-customers, as the other companies, would therefore become richer and richer in infinite simulations.

- *Speculators are not cost efficient:* Our speculators never earn a large amount of money. For instance, they own a maximum average amount of 156.4 shared among 5 wood speculators, but the corresponding large standard-deviation of 237.9 indicates that several of these speculators have negative funds in Week 500. Yet, speculators, like all agents, start with 60 units of money.

This remark about the poor cost efficiency of speculators is rather counter-intuitive! In addition, speculators are the only agents in our simulation which loose money. And this latter point would not change if simulations were longer, because all simulations end with a market at the equilibrium in which they cannot trade, thus, earn money.

6.4 Related Work

In this section, we outline our reproduction of Steiglitz and his colleagues' model [5, 10, 11]. As for our supply chain simulation, we use the parameters in [11]. More precisely, the speculators used here are exactly the same as those used in our supply chain, and the trading agents share most of their code (in particular, their bidding strategy and their valuation policy) with our supply chain agents.

Figure 6.7 shows that the speculators are as efficient in stabilising the price in our supply chain as in a single market. Subfigures 6.7(a) and 6.7(b) show that the speculators have a stabilising effect, and the table in Subfigure 6.7(c) that the speculators lose money when they trade (they all begin with 60 units of money and, on average, end up with a negative amount). As in our supply chain simulation, traders should pay the speculators to bring stability. Speculation does not seem to increase greatly the average welfare of traders (i.e the average final amount of money only increases from 616.4 to 664.0, which is not significant since only one simulation run is considered), but money is much better shared between the traders (i.e. the standard-deviation of final funds decreases by two orders of magnitude from 1,395.7 to 67.6).

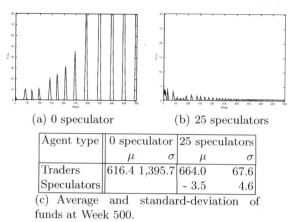

(a) 0 speculator   (b) 25 speculators

Agent type	0 speculator		25 speculators	
	μ	σ	μ	σ
Traders	616.4	1,395.7	664.0	67.6
Speculators			- 3.5	4.6

(c) Average and standard-deviation of funds at Week 500.

Fig. 6.7. Reproduction of Steiglitz *et al.*'s experiments [5, 10, 11] (all subfigures have the same scale as Figure 6.3).

6.5 Discussion

Since speculators lose money when they stabilize the supply chain, then companies should pay them for their service of stabilisation. The question is then who should pay for this stability? Since price stability can be seen as a common good (or, at least, serice), a prisonner's dilemma may occur, that is, everyone would like money be given to speculators, but nobody wishes to pay because all companies prefer all other companies pay but not themselves.

Another point to consider is the second type of speculators proposed by Steiglitz *et al.* [10, 11], namely DER speculators (for derivatives). These speculators estimate the second derivative of the price slope. They sell when the slope of the price is increasing and buy when the slope is decreasing. Perhaps these agents can both stabilize our supply chain and be cost efficient.

6.6 Conclusion

This paper has proposed to use some sort of speculators to reduce the bullwhip effect, which is the amplification of order variability in a supply chain. Reducing this effect is important because it costs money to the companies due to higher inventory levels and because the bullwhip decreases supply chain agility. To our knowledge, using speculators as a solution to the bullwhip effect is a new approach. We have explored this idea by means of a computer simulation of a supply chain, in order to quantify the impacts of speculation on the price fluctuations caused by the bullwhip effect.

This is still a preliminary work, but three conclusions can be drawn. The first one is that adding a few speculators stabilizes the price in the market of the supply chain in which they are introduced. The second conclusion is that price bubbles may appear, and this seems occur when too many speculators are used. Finally, our speculators are not cost efficient and lose their initial level of money to an amount near zero.

As future work, we intend to run more experiments to confirm the conclusions in this paper. We also plan to compare different types of speculation (financial vs. stock) and study the potential benefits of futures contracts in supply chains. An application of these ideas concerns the effective control of online marketplaces, such as those emerging for scientific and commercial computational resources known as the GRID. If the presence of speculators in markets is shown to reduce the bullwhip effect, then designers of GRID distribution chains may engineer their systems to include artificial speculators.[4]

References

[1] Carlsson, C. and Fullér, R. (2000). Soft computing and the bullwhip effect. *Economics and Complexity*, 2(3).

[2] Kaldor, N. (1960). *Essays on Economic Stability and Growth*, chapter Speculation and economic stability, pages 17–58. Gerald Duckworth, London, UK.

[3] Lee, H. L., Padmanabhan, V., and Whang, S. (1997). Information distortion in a supply chain: The bullwhip effect. *Management Science*, 43(4):546–558.

[4] Marengo, L. and Tordjman, H. (1996). Speculation, heterogeneity and learning: A simulation model of exchange rates dynamics. *Kyklos*, 49(3):407–438.

[5] Mizuta, H., Steiglitz, K., and Lirov, E. (2003). Effects of price signal choices on market stability. *Journal of Economic Behavior and Organization*, 52:235–251.

[6] Moyaux, T., Chaib-draa, B., and D'Amours, S. (2006). Information sharing as a coordination mechanism for reducing the bullwhip effect in a supply chain. *IEEE Trans. on Systems, Man, and Cybernetics*. (to appear).

[7] Palmer, R. G., Arthur, W. B., Holland, J. H., LeBaron, B., and Tayler, P. (1994). Artificial economic life: a simple model of a stockmarket. *Physica D*, 75(1-3):264–274.

[8] Phelps, S. (2006). Web site for JASA (Java Auction Simulator API). http://www.csc.liv.ac.uk/~sphelps/jasa/ (accessed February 23, 2006).

[9] Simchi-Levi, D., Kaminsky, P., and Simchi-Levi, E. (2000). *Designing and managing the supply chain*. McGraw-Hill Higher Education.

---

[4] This research was undertaken as part of the EPSRC funded project on Market-Based Control (GR/T10664/01).

[10] Steiglitz, K., Honig, M. L., and Cohen, L. M. (1996). A computational
market model based on individual action. In Clearwater, S. H., editor,
Market Based Control, pages 1–27. World Scientific: Singapore.

[11] Steiglitz, K. and Shapiro, D. (1998). Simulating the madness of crowds:
Price bubbles in an auction-mediated robot market. *Comput. Economics*,
12:35–59.

[12] Supply Chain Council (2001). Overview of the SCOR model v5.0.
http://www.supplychainworld.org/WebCast/SCOR50_overview.ppt
(accessed Dec. 2001).

Part III

Firm-Consumer Dynamics

Co-evolutionary Market Dynamics in a Peaked Resource Space

César García-Díaz[1] and Arjen van Witteloostuijn[1,2]

[1] Department of International Economics and Business,University of Groningen,
The Netherlands c.e.garcia.diaz@rug.nl
[2] Durham Business School, University of Durham, UK
a.van.witteloostuijn@rug.nl

7.1 Introduction

Market processes involve simultaneous interactions among firms and consumers. Firms target segments with abundant "resources" (i.e. high purchasing power or just demand), while consumers search for firms' offers that best
match their preferences. We explore implications of this dual interaction in
which large-scale and small-scale firms compete in an initially established
peaked resource space with a center, where resources (i.e. consumers) at the
starting date (time = 0) are assumed to be more abundant in the central
region than in the periphery. The resource space represents the distribution
of consumers along a one-dimensional set of taste preferences. We explore
the implications for the evolution of market structure (number of firms and
market concentration) under this rendering by means of an agent-based simulation model. We derive two patterns of results. First, when firms move to
the best spots in the market and consumer mobility along the taste positions
is prohibited (i.e. the resource space shape is constant over time), the market
exhibits high concentration with first increasing and then declining number
of firms, although such number remains relatively high. Second, when consumers are allowed to update their taste preferences while searching for a
best match, the increased concentration pattern may be broken down and a
higher small-scale firm proliferation might be observed. In addition, consumer
mobility reinforces the large-scale firms' space contraction.

7.2 Background

In recent years, co-evolutionary processes of markets and organizations have
started to draw attention of social scientists, in both theoretical and empirical
domains. Researchers have addressed co-evolutionary issues related to, among
others, empirical studies design (Lewin and Volberda 1999), price dispersion

effects (Kirman and Vriend 2001), joint ventures (Inkpen and Currall 2004), strategic alliances (Koza and Lewin 1998), and market dominance (Harrington and Chang 2005).

Organization Ecology (Hannan and Freeman 1989) has shown that, in many instances, consumers are allocated according to a unimodal distribution in a space of taste preferences (Carroll et al. 2002; Boone and Witteloostuijn 2004; Witteloostuijn and Boone 2006). Market configurations that emerge in such resource peaked spaces (Carroll and Hannan 1995) have been extensively studied empirically in a wide variety of industries such as newspapers (Carroll 1985; Boone et al. 2002, 2004), breweries (Swaminathan 1998; Carroll and Swaminathan 2000), automobile manufacturers (Dobrev et al. 2001), wineries (Swaminathan 1995, 2001) and audit firms (Boone et al. 2000). In such setting, where peaked spaces constitute attractive places for reaping scale economies, it has been observed that specialist organizations, those that serve a narrow niche (a small set of taste preferences), proliferate as market concentration rises. It has also been argued that specialists proliferate not only due to the lack of capabilities of generalist firms to reach the extremes of the resource space, but also due to the ability of specialists to exploit both unused peripheral resources as well as the importance of customer identity and self-expression in such marginal niches. Carroll and Hannan (1995) argue that mature markets sometimes tend to reflect the flattening of the resource space, as a by-product of the ability of specialist organizations to open up new niches (Swaminathan 1998). This implies a co-evolutionary process in which firms compete, take advantage of scale economies, grow large and consolidate by targeting the best spots in the resource space, while consumers refine their tastes by moving towards the firm that best matches their preference: large-scale firms might find it costly to cover the whole space and many consumers might find it more attractive to be served by specialized firms at the periphery.

An aspect that may well be critical here is the extent of consumer mobility. Consumer search models have been extensively studied (see for instance, Stahl 1989) in the Industrial Organization tradition, and the effects of search and switching on industry performance have been clearly acknowledged (Waterson 2003). For instance, price competition in a single-product oligopoly or perfect contestability context is known to be extremely tough if consumers are perfectly mobile. Only then, after all, do clients move from one firm to the other if the latter offers a price that is lower than the former's even when this difference is infinitely small only.

Following the role that computational and mathematical modeling has played in studying such industry evolution processes in Organizational Ecology (Lomi and Larsen 1996, 1998; Péli and Nooteboom 1999; Barron 1999, 2001; Harrison, 2004; Lomi et al 2005) we explore the firm-consumer dual dynamics in an agent-based model under a set of specific conditions: a peaked resource space with heterogeneous consumer preferences, and a changing set of large-scale and small-scale firms. Our key contribution is that we analyze the impact of different degrees of consumer mobility on market structure evolution. Following ecological theories, we formulate our model with exogenous entry and endogenous exit, but the novelty of our representation is that we add a reciprocal interplay between firms and consumers (Lewin et al. 2004) as a driver of market structure evolution to the traditional ecological view where markets are mainly considered to be shaped by entry and exit rates of firms (Hannan and Freeman 1989; Carroll and Hannan 2000).

7.3 Summary of the Model

Next, we introduce an agent-based model with an initial peaked demand distribution, scale economies and consumer taste heterogeneity. Demand is taken to be distributed along 100 different taste preferences using a Beta distribution with parameters $\alpha = \beta = 3$. Firms are of two types (large-scale and small-scale), enter the market at some taste position and gradually move towards the most abundant spots (e.g. the "peak" or market center) as they grow. Consumers update their taste preferences according to either the closest match according to their current taste or the highest expected utility.

7.3.1 Firm Behavior

The model starts with one single firm. Firm entry to the market is drawn from a negative binomial distribution, which depends on a density-dependent rate (Harrison 2004)[3]. Such density-dependent mechanism is consistent with empirical findings regarding organizational founding (Hannan and Carroll 1992; Barron 1999): we consider a process with an arrival rate represented by $\lambda(t)$ $= \exp[\delta_0 + \delta_1 N(t) + \delta_2 N(t)^2]$, where $N(t)$ is firm density at time t, with t= 1,2,...,T. Parameter values for δ_0, δ_1 and δ_2 are calibrated in the same fashion as done in previous density-dependence models (see for instance, Harrison 2004). In addition, there is a distribution algorithm of entrants over the one-dimensional resource space based on two facts: a) the probability of founding a large scale is 1 at $N = 0$, and, b) such probability is a monotonic decreasing function of the total industry output: as the industry output approaches the market's carrying capacity[4], the probability of founding a large-scale firm decreases (see Carroll and Hannan 1995). Consequently, the probability of founding a small-scale firm is just the complement.

The cost function of a firm has two components, one related to production cost, and the other one related to niche-width costs. Thus, for firm i at time t, total costs at time t are represented by the production costs, $C_P^i(t)$, and niche-width related costs $C_{NW}^i(t)$:

---

[3] Hereafter, following the Organizational Ecology tradition, we define density as the number of firms in the market.

[4] In Organizational Ecology, carrying capacity is defined as the maximum number of firms (i.e., maximum density N) that can operate viably in the market.

$$C^i(t) = C_P^i(t) + C_{NW}^i(t) \tag{7.1}$$

The production function corresponds to a classic Cobb-Douglas function with two production factors: a quantity-independent (fixed) one and a quantity-dependent (variable) one. Values of the Cobb-Douglas function were chosen assuming that the long-run average cost curve is downward sloping (in order to reflect scale economies) and with a minimum (normalized) value of 1. Niche-width related costs appear as firm expands horizontally along the taste preference axis, and reflect the complexity (i.e. "scope" diseconomies) of handling a large number of different taste preferences. We refer to "niche" as the set of taste positions where the firm sells product. Each niche has a center that is updated as the firm moves. A firm moves in the direction in which "resources" are more abundant, so they can benefit from reaping scale effects and reducing prices. We define $w_i^u(t)$, $w_i^u(t)$ as the upper and lower niche limits of firm i, and NWC as the niche-width cost coefficient. The niche-width costs are:

$$C_{NW}^i(t) = NWC * (w_i^u(t) - w_i^l(t) \tag{7.2}$$

7.3.2 Consumer Behavior

Consumers are distributed according to a unimodal one-dimensional resource space with a market center. As mentioned above, demand is distributed among n taste positions (n=100). Each consumer buys only one product each time period. Each taste position k is characterized by a number of consumers b_k. Consumer j at taste position k has a utility function defined by:

$$U_{j,k}(i,t) = B_j(i,t) - P_i(t); \quad j = 1, 2, ..., b_k; \ k = 1, 2, ..., n \tag{7.3}$$

The term $B_j(i,t)$ is the "benefit" consumer j receives (e.g. product functionality) at time t, and $P_i(t)$ is the price she or he pays to firm i. We assume that the benefit for acquiring a product decreases with taste distance (Hotelling 1929). We define $B_j(i,t)$ as:

$$B_j(i,t) = B_o - [\gamma \frac{\|p_i(t) - k\|}{n} + \varepsilon_{ij}] \tag{7.4}$$

The term Bo is a constant, $p_i(t)$ is firm i's niche center, $\|p_i(t)-k\|$ is the distance between the firm's niche center an the taste position, γ is constant and ε_{ij} an error term that represents the inability of consumer j to exactly evaluate "product dissimilarity" respect to her or his own taste. The term ε_{ij} is assumed to be normally distributed with mean = 0 and standard deviation = 0.05. When buying, each consumer at position k maximizes her or his utility according to a utility participation constrain U_o. U_o is set up as a markup (20%) on the maximum value of the long-run average cost curve.

7.3.3 Model Dynamics

Prices are initially set up by estimating the expected additional quantity firms will obtain in the next time period. Let us consider a firm (e.g. firm i) that enter the market at an empty slot k at time t. Let us define $Q_i(t)$ as the quantity firm i expects to get, $U_i(t)$ as the utility that firm i offers to consumers at position k, b_k. Then,

$$Q_i(t) = b_k P(U_i(t) > U_o) \qquad (7.5)$$

If firm i enters an occupied slot, the entrant follows exactly the same procedure, except that the calculation of Q_i includes information from the N_k competing firms already offering product at that taste position:

$$Q_i(t) = b_k P(U_i(t) > U_o) \prod_{j=1}^{N_k} P(U_i(t) > U_j(t-1)) \qquad (7.6)$$

After competition, niche limits $w_i^l(t)$ and $w_i^u(t)$ are adjusted accordingly depending on lost or gained taste positions. Firms also update their niche center $p_i(t)$. Firms engage in both vertical and horizontal expansion. Expansion is assumed to be dependent on expected incremental consumer target and an expansion probability. Firms start with a price that depends on others' prices, but update its level to a markup price depending on future scale economies gains. The markup reflects the opportunity cost for a firm in the industry. Firm stay in the market as long as they have non-negative profits. As mentioned above, firms are of two types: large-scale and small-scale. A large-scale firm is calibrated so that it catches between 1/2 and 2/3 parts of the whole resource space in the absence of competition. For this aim, the values of the coefficients for NWC, γ and the Cobb-Douglas parameters are jointly calibrated (NWC = 200, γ = 10). In addition, small-scale firms are calibrated at a lower point in the long-run average cost curve, in order to reflect lower scale advantages *vis-à-vis* large-scale firms. For small-scale firms, a set of different values for the Cobb-Douglas function is used in the simulation trials. For convenience, we assume that positions never disappear for lack of demand due to the mobility process. That is, taste positions always have some demand to offer. So, both total demand and the total number of positions remain constant throughout the simulation experiments. We assume that the minimum demand a taste position has is one consumer.

7.4 Simulation Experiments and Results

Each simulation was run for T = 400 time periods. Simulation trials were run for two different values for the firm expansion probability, which were taken from the approximate extremes of its calibrated value range (high = 0.15

and low = 0.05). Accordingly, we experimented with three different values of long-run average cost curve points, denoting three different cost curves for the small-scale firms, the two rather extreme values of the calibrated range and a middle one (low = 5, moderate = 10 and high = 20). Therefore, the total number of parameter combinations is six. Each combination was run five times and the results were averaged. We performed 6x5 = 30 simulation runs for each experiment. We run three different experiments, which are explained below, for a total of 90 simulation runs.

7.4.1 Baseline Model (Without Consumer Mobility)

As our baseline, we assume that the resource space shape remains constant over time. That is, consumers are not mobile at all. Results show increasing concentration (C_4concentration ratio) coupled with an initially increasing and later a declining density. Few large-scale firms take over the market center while small-scale firms move to the peripheral areas. This somehow reflects a market partitioning similar to that found in Organizational Ecology's resource-partitioning theory (Carroll 1985; Carroll et al. 2002), although overall density declines after reaching a peak, which does not follow from the theory's original prediction. According to the results presented in Fig. 7.1 (dashed lines represent sample runs and solid lines represent average behavior), we observed that the organizational density, on average, tends to slightly decline below 150 firms. Market concentration shows an increasing trend with a value above 70 per cent after the 400 time periods.

7.4.2 Consumer Mobility According to Closest Matched Taste

Next, we assume that consumers move into the direction where they expect to find firms that match closer with their current taste. In this case, the mobility decision does not involve prices, although consumers use Eqs. 7.3 and 7.4 to assess the best option when purchasing. Empirical evidence in the U.S. brewery and wine industry show that consumer might be inclined to move to peripheral taste preferences heavily based on identity reasons, no matter whether or not premium prices appear (Carroll and Swaminathan 2000; Swaminathan 2001). Consumers inspect adjacent taste positions (one to the left, one to the right) and move to the position where a closest-to-own-taste product is offered (i.e. the offering where the shortest distance between the firm's niche center and the current consumer's taste is found), according to a constant mobility rate of θ per time period (θ=5%) per taste position[5].

---

[5] Alternatively, we could have used a mobility probability (per consumer), consistent with the consumer agent's buying process, instead of a mobility rate per taste position. There are two reasons for not doing so: first, the mobility process might belong to some aggregate level that should not be modeled at the individual level without a rather developed set of consumer interactions (e.g., network externali-

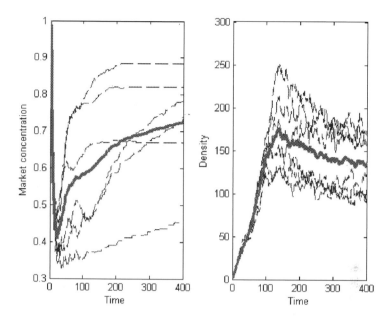

Fig. 7.1. Consumer immobility

Results reveal that the scale effect of large-scale firms does not diminish with consumer mobility, since the average trend of market concentration (C_4 ratio) is still associated with an increasing trend to a level above 70 per cent at the end of the simulation horizon. However, the mobility process seems to affect the way firms proliferate, since density tends to stabilize at a point around 200 firms, well above the level in the case without consumer mobility. Taking into account that simulated data reveals a very low number of large-scale firms at time = 400, this result reflects a positive effect on small-scale firms' proliferation. This pattern of results is reproduced in Fig. 7.2.

7.4.3 Consumer Mobility According to Highest Expected Utility

We assume now that consumers move into the direction of higher utility spots. This is, consumers inspect others' utility offerings in adjacent taste positions and decide to move according to a constant mobility rate θ per time period ($\theta=5\%$) per taste position. Unlike the previous experiments, in some cases market now evolves toward a fragmented structure, reflecting rather low levels of concentration. This is clear from Fig. 7.3.

---

ties), which is beyond the scope of this work. Second, a mobility probability will imply a higher computing intensity, which is an unnecessary complication not needed to derive the results we want to report here.

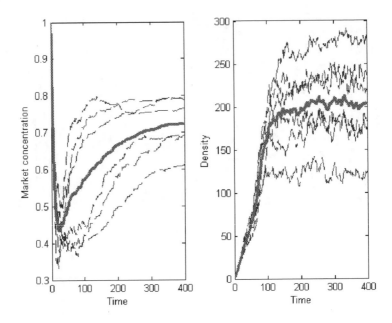

Fig. 7.2. Match-improving consumer mobility

In any case, concentration does not show an increasing trend and falls well below the level observed in the two previous experiments, revealing that the large-scale firms' advantage is weakened. It is worth noticing that, for the cases with low market concentration, the market is divided among rather similar and small firms (fragmentation). This case reflects that the proliferation of small-scale firms is due in part to the fracture of scale advantages, since the dominance of large-scale firms might be lost.

7.4.4 Effects on Large-Scale Firms Space

Organizational ecologists have argued that, in resource-partitioning processes, the total space own by generalist organizations decreases as concentration rises (Carroll and Hannan 2000; Carroll et al. 2002). We now investigate the effects on space contraction of large-scale firms under our three consumer mobility scenarios. We define the large-scale firms' total space as the aggregated number of taste positions that such firms serve.

In absence of consumer mobility (experiment 1), it appears that whether or not the large-scale firms' space gets reduced depends on the value of the capacity of expansion of the small-scale firms. Our interpretation is that, as such capacity increases, the more likely it becomes that such small-scale firms appropriate space from large-scale firms, since the former are in better position to make more attractive offerings to peripheral consumers. However, results

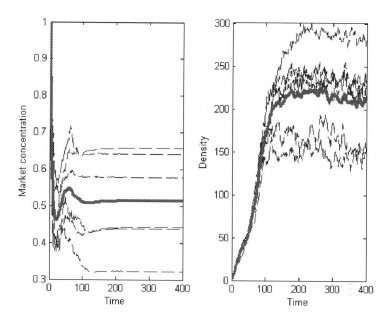

Fig. 7.3. Utility-maximizing consumer mobility

from experiments 2 and 3 reveal that consumer mobility helps to reinforce such space contraction. The average large-scale firms' space clearly contracts over time, as can be seen from Fig. 7.4.

7.5 Conclusion

It has been shown how firm-consumer dual dynamics might greatly affect the evolution of market structures. The key finding is that consumer mobility might diminish the power of scale effects. Industry evolution theories still focus heavily on the supply side, without incorporating consumer effects and firm-buyer dynamics. The presented results differ from Harrington and Chang's (2005), since they argue that firm-consumer dual dynamics might lead to market dominance. However, the context and assumptions of both models differ greatly (Harrington and Chang's model is about two firms adapting and searching for innovations to match consumers' attributes in absence of price mechanisms). In addition, further research along these lines may try to explain the flattening of the resource space, as argued by organizational ecologists. First, it is shown under an agent-based framework that small-scale firms may enhance their proliferation in a market dominated by large-scale firms if the former properly exploit identity-related tastes. Second, consumer mobility can also weaken scale advantages, and further stimulate small-scale

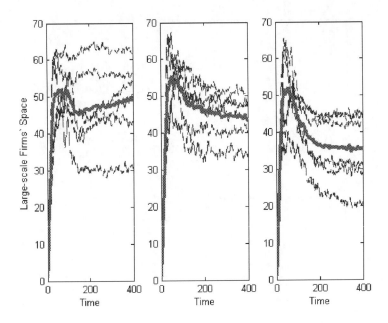

Fig. 7.4. Large-scale firms' total space for experiments 1 (left), 2 (center) and 3 (right)

firm proliferation. As seen in the first experiment, scale economies might not be enough to explain resource-partitioning's resource-release effects, and some form of consumer mobility would be needed to explain it fully. This suggests, again, that the "supply-side" resource release outcome might need a complementary "demand-side" consumer mobility explanation.

Acknowledgement

Part of this work was carried out when the first author was visiting scholar at the University of Texas at Dallas. J.R. Harrison's support is highly appreciated. We are grateful to S.A. Delre, G. Péli and three anonymous reviewers for their valuable insights about this work. We also thank the SOM research school at the University of Groningen and The Netherlands Organization for Scientific Research (NWO) for financial support. The usual disclaimer applies.

References

[1] Barron D (1999) The structuring of organizational populations. American Sociological Review 64(3): 421-445

[2] Barron D (2001) Simulating the dynamics of organizational populations. In: Lomi A, Larsen E (eds) Dynamics of organizations: Computational modeling and organization theories. Cambridge, MA: MIT Press, pp. 209-242

[3] Boone C, Wittteloostuijn A van (2004) A unified theory of market partitioning: an integration of resource-partitioning and sunk cost theories. Industrial and Corporate Change 13: 701-725

[4] Boone C, Bröcheler V, Carroll G (2000) Custom service: Application and test of resource-partitioning theory among Dutch auditing firms from 1896 to 1992. Organization Studies 21(2): 355-381

[5] Boone C, Carroll G, Witteloostuijn A van (2002) Resource distributions and market partitioning: dutch daily newspaper organizations, 1968 to 1994. American Sociological Review (67): 408- 31

[6] Boone C, Carroll G, Witteloostuijn A van (2004) Size, differentiation and the performance of Dutch daily newspapers. Industrial and Corporate Change 13(1): 117-148

[7] Carroll G (1985) Concentration and specialization: dynamics of niche width in populations of organizations. American Journal of Sociology 90 (6): 1262-1283

[8] Carroll G, Hannan M (1995) Resource partitioning. In Carroll G, Hannan M (eds) Organizations in industry: strategy, structure and selection. Oxford University Press, pp. 215-221

[9] Carroll G, Hannan M (2000) The demography of corporation and industries. Princeton University Press

[10] Carroll G, Swaminathan A (2000) Why the microbrewery movement? Organizational dynamics of resource partitioning in the U.S. industry. American Journal of Sociology 106(3): 715-62

[11] Carroll G, Dobrev S, Swaminathan A (2002) Organizational processes of resource partitioning. In: Kramen R, Staw B (eds) Research in organizational behavior, New York: JAI/Elsevier

[12] Dobrev S, Kim T, Hannan M (2001) Dynamics of niche width and resource partitioning. American Journal of Sociology 106: 1299-1337

[13] Hannan M, Freeman J (1989) Organizational ecology. Harvard University Press

[14] Harrington J, Chang M (2005) Co-Evolution of firms and consumers and the implications for market dominance. Journal of Economic Dynamics and Control 29: 245-276

[15] Harrison JR (2004) Models of growth in organizational ecology: a simulation assessment. Industrial and Corporate Change 13(1): 243-261

[16] Hotelling H (1929). Stability in competition. Economic Journal 39(153): 41-57

[17] Inkpen, A, Currall S (2004) Coevolution of trust, Control, and learning in joint ventures. Organization Science 15(5): 586–599

[18] Kirman A, Vriend N (2001) Evolving market structure: An ACE model of price dispersion and loyalty. Journal of Economic Dynamics and Control 25: 459-502

[19] Koza M, Lewin A (1998) The coevolution of strategic alliances. Organization Science 9:255–264

[20] Lewin A, Volberda H (1999) Prolegomena on coevolution: A framework for research on strategy and new organizational forms. Organization Science 10(5): 519-534

[21] Lewin A, Weigelt C, Emery J (2004) Adaptation and selection in strategy and change: Perspectives on strategic change in organizations. In: Poole MS, Van de Ven A (eds) Handbook of Organizational Change and Innovation. Oxford University Press, pp 108-160

[22] Lomi A, Larsen E (1996) Interacting locally and evolving globally: a computational approach to the dynamics of organizational populations. Academy of Management Journal 39(4): 1287-1321

[23] Lomi A, Larsen E (1998) Density delay and organizational survival: computational models and empirical comparisons. Computational and Mathematical Organization Theory 3(4): 219-247

[24] Lomi A, Larsen E, Freeman J (2005) Things change: dynamic resource constrains and system-dependent selection in the evolution of organizational populations. Management Science 51(6): 882-903

[25] Péli G, Nooteboom B (1999) Market partitioning and the geometry of the resource space. American Journal of Sociology (104): 1132-1153

[26] Stahl D (1989) Oligopolistic pricing with sequential search costs. American Economic Review 79(4): 700-712

[27] Swaminathan A (1995) The proliferation of specialists organizations in the American wine industry, 1941-1990. Administrative Science Quarterly 40: 653-680

[28] Swaminathan A (1998) Entry into new market segments in mature industries: endogenous and exogenous segmentation in the U.S. brewing industry. Strategic Management Journal 19: 389-404

[29] Swaminathan A (2001) Resource partitioning and the evolution of specialists organizations: the role of location and identity in the U.S. wine industry. Academy of Management Journal 44(6): 1169-1185

[30] Waterson M (2003) The role of consumers in competition and competition policy. International Journal of Industrial Organization 21: 129-150

[31] Witteloostuijn A van, Boone C (2006) A resource-based theory of market structure and organizational form. Academy of Management Review 31(2): 409-426

E-Consumers' Search and Emerging Structure of Web-Sites Coalitions

Jacques Laye[1], Maximilien Laye[2], Charis Lina[3], and Hervé Tanguy[2]

[1] LEF Inra Sae2/Engref, Nancy, France, `laye@nancy-engref.inra.fr`
[2] Laboratoire d'Économétrie de l'École Polytechnique, Paris, France
[3] University of Crete, Greece

Summary. This article summarizes the main results we obtained in an agent-based extension of an industrial organization analytical model[4] for studying emerging coalition structures in electronic markets.

8.1 Introduction

As Axtell (2000) points out, "an agent-based computational model is valuable to study the non-equilibrium dynamics, in which structure is perpetually born, growing and perishing", which is the case of the constantly evolving Internet landscape. We develop an agent-based simulation model in order to study the forces that drive the Web-sites aggregation phenomena, following the principles described in the related literature (see for instance the work of d'Inverno *et al.* (1997), Wooldridge *et al.* (2000), and Kinny (1999, 2001)). We study a virtual environment consisting of Internet consumers, online sellers (*i.e.* B-to-C Web-sites) and rules of interaction among these economic agents. Most of the agent-based approaches that focus on the Internet economy consider that the most important factor of the evolution of the Internet landscape are network effects and increasing returns, see for instance Lina (2003). Our approach differs from these contributions in two ways. Firstly, our analysis does not take into consideration these characteristics of the Internet landscape (information feedback effects, sites' underlying network, presence of Web-communities, *etc.*). In other words, the Web here is an alternative channel of distribution and not an entertainment good. We focus on the aggregation strategy of pure merchant sites (coalitions), trying to capture consumers only willing to buy online by reducing their search cost. The reduction of search costs is considered to be the result of the coordinated efforts of the coalesced sites to develop more efficient search tools in order to facilitate the finding of the goods closest to the tastes of Internet consumers (mutual electronic linkage among sites or through the development of specialized search engines).

---

[4] Laye (2003), Laye and Tanguy (2005).

This defines the notion of coalition we are interested in: reduction of search cost as a possibility to increase the expected demand rather than price coordination or merging. We base our multi-agent model over a theoretical one (fully presented in Laye (2003) and Laye and Tanguy (2005)) that is related to the literature of Industrial Organization, which allows us to take into account economic behaviors (agents are maximizing their utility) rather than pure behavioral rules. Agents, mimicking the behavior of Web sites and consumers, interact through two basic processes: the consumers' search process and the coalition formation process. Extending the analytical model by introducing dynamics and heterogeneity on the agents' characteristics and behavior enables us to study more systematically the emergent coalition structures in a more complex, and thus more realistic, environment. The agent-based model incorporates a dynamic process of coalition formation where coalitions can be formed in parallel. Contrary to the analytical model, the number of sites and consumers in the simulated market grows over time. A process of market entry/exit is also added. Moreover, sites have the ability to dynamically adjust their price. In various configurations of the agent-based model, we look at the economical results in terms of *(i)* degree of differentiation of the coalesced sites , *(ii)* the number and size of the formed coalitions.

8.2 An Example of Search Within Coalitions

Let us describe the search procedure of a traveler willing to book online in a 5-star hotel in Paris for a conference. 5-stars hotels are mainly located in a few neighborhoods of Paris (Champs-Elysées, Opéra, *etc.*) that will be the criterion of differentiation for the traveler (which one is closer to the conference center?). The traveler anticipates a 5-star hotel to cost 300 Euros, but discovering a lower price could be a motivation to accept a more distant hotel, if the difference in price counterbalances the adaptation cost. By typing a request in a search engine (such as google.com) a great number of sites are found, leading to a loss of time and energy (search cost) in order to get the information about price and location of one hotel. If the hotel is "close enough" to the consumer's preference, the transaction can take place, otherwise it is preferable to perform a new search, depending on the characteristics of the current hotel, on the priors about future hotels to discover, and on the search cost. The traveler gets sites of hotels (Le Crillon, Ritz, *etc.*), but also portal sites ("coalitions") that reduce the search cost of the consumer thanks to specialized search engines: by sorting the results of the site by stars, it is possible to directly find with a lower search cost many 5-star hotels (2 in parishotel-reservation.com, 4 in paris.book-online.org, 5 in hotel-paris-tobook.com and hotelclub.org, 6 in 0800paris-hotels.com). In such a site, hotels increase the probability of being visited, thereby increasing the expected demand, but they remain independent in their price policy. We empirically observe that inside a coalition hotels belong to different neighborhoods. For instance in 0800paris-hotels.com, the 6 luxury hotels proposed by the site are located in 6 different districts of Paris. Our goal is to find the rationale behind the choice of a highly differentiated partner rather than another coalition structure.

8.3 Analytical Model

As in the model of Bakos (1997), we consider a circular model of spatial differentiation that represents both the characteristics of the differentiated goods offered by Web-sites and the tastes of consumers. More precisely, we consider a market with a continuum of Internet consumers and m B-to-C sites. m is supposed to be common knowledge. Each site j sells a unique good at price p_j and the characteristics x_j of the goods are differentiated along the unit circle. The tastes x_i of the consumers are heterogeneous and uniformly distributed along the same circle. By buying a unit of good that does not match exactly with their preference, consumers incur an adaptation cost t per unit of distance ($t > 0$) between their location (*i.e.* their preferred product) and the location on the circle (*i.e.* the good offered) of the site chosen for the transaction. Therefore, the utility function if consumer i buys a unit offered by site j is: $U(i,j) = r - p_j - t|x_i - x_j|$, where r is the reservation utility of each consumer.

Consumers' Search Procedure

As in Gabszewicz and Garela (1986) or Bakos (1997), we suppose that sites compete in price and that the population of consumers is imperfectly informed. Consumer i acquires information on the location and the price of one of the m sites of the electronic market by incurring a constant search cost $c > 0$. We consider this search cost to be both the cost associated with the discovery of the site on the Web, for example through a search engine, and the cost of visiting the site to find out about its characteristics: sell price S, and distance D. The utility of the consumer in case of a transaction is $U(S,D) = r - S - tD$. If the consumer decides to search further and finds another site located at distance x and with price p, the utility is $U(p,x) = r - p - tx$. Thus, $U(x,p) - U(S,D))+ = (S + tD - xt - p)^+$ represents the increase of utility for the consumer if $U(x,p) - U(S,D) > 0$ (otherwise it is 0). We suppose that the consumers are risk neutral. The calculation of the expected gain in utility based on the priors on the distributions of sites' locations and prices allows the consumer to decide on the opportunity to continue the search procedure.

Consumers' Priors

Concerning the priors on prices, the consumers believe that at equilibrium all sites choose the same price p^*. More precisely, the distribution of prices is such that $f(p) = 1$ if $p = p^*$, and $f(p) = 0$ otherwise. Concerning the priors on locations, consumers believe that sites locate according to a uniform distribution over the unit circle. We also suppose that consumers find sites according to a random trial with replacement. These assumptions are related to the fact that consumers are considered to not change their priors on the distributions of locations or prices after finding each site.

Stopping Rule

The expected gain in utility obtained in this case is: $g(S, D) = \int_{x=0}^{1}(\int_{\Re}(S + tD - xt - p)^{+}f(p)dp)dx$. Given the priors on the locations, we find like in Bakos (1997) that $g(S, D) = (S+tD-p^*)2/t$. Next, consumers compare their expected gain in utility with the search cost c. If $g(S, D) > c$, a consumer will prefer to continue the search. If $g(S, D) < c$, a consumer will choose to buy a unit of the good located at a distance D and at price S. At equilibrium with rational expectations for the consumers, $S = p^*$. For each consumer i located in x_i, we have that $g(p^*, D) < c$ on the interval $[x_i - L, x_i + L]$, where $L = \sqrt{c/t}$. Consequently, if the consumer discovers a site at a distance smaller than L, the transaction will take place. Symmetrically, from the point of view of a site, the more distant potential client is located at distance L. We obtain an interval of length $2L$ around any site, which will be referred to as "natural territory". The natural territory of a site corresponds to the interval around its location in which consumers stop their search and buy from this site if they find it.

Coalitions

Next, we consider that sites have the possibility of forming coalitions and that this leads to a reduction of the search cost for a consumer that visits a site within a coalition. A consumer that has only incurred the search cost c for a search on the entire Web can visit other sites by incurring a lower cost $c' < c$ within the coalition. We consider that there are 4 sites located according to the principle of maximum differentiation and selling at price p^*, which is also the price anticipated by the consumers. We restrict the study in terms of length of natural territories by supposing that $\underline{L} < L < \overline{L}$ such that no consumer is priced out of the market and the natural territory of a site only intersects with those of its neighbors. This setting is the minimal setting required to differentiate coalition structures. Indeed, a site willing to coalesce can choose two kind of partners defining two different categories of coalitions. A coalition will be called "connex" if the natural territories of its members intersect, otherwise the coalition will be called "non-connex". As shown in Figure 8.1, for $\underline{L} < L < \overline{L}$, a coalition is connex if its members are located consecutively on the circle (little differentiation between sites 1 and 2), and non-connex otherwise (high differentiation between sites 1 and 3).

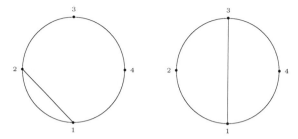

Fig. 8.1. Connex and non-connex coalition structures

Results of the Analytical Model

Two main questions motivate this analytical model: absent price coordination, what is the economic rationale behind the coalition formation of specialized B-to-C Web-sites and what is the more preferable type of coalition: those involving sites selling little differentiated products (connex) or highly differentiated ones (non-connex)? As it is extensively described in Laye (2003) and Laye and Tanguy (2005), the static comparative of the two coalition structures shows that a site willing to coalesce has more incentives to choose a non-connex partner. It is shown that for both coalition structures (connex and non-connex), coalesced sites have an incentive to lower their prices from the price obtained without coalitions in order to increase their natural territory. The opposite tendency is observed for the non-coalesced sites: they increase their price in order to decrease the length of this territory. Furthermore, the non-connex coalition is more aggressive than the connex one. The fact that non-connex partners decrease more their prices than if they were in a connex coalition shows that it is not the increase in the competition between the coalesced sites that drives the price decrease. Decreasing the price reflects only the opportunity to gain market share from the non-coalesced sites. Given that the existence of search costs that are independent from the adaptation costs is a specific characteristic of Internet distribution, this analytical model can enlighten the discussions on the emerging structures of B-to-C coalitions, as part of the evolving Internet landscape. However, we are interested in the way in which more complex dynamics on the population of agents (sites and consumers) and on the coalition formation can influence the mechanisms captured by the analytical model, which is a motivation for the agent-based model presented in the next sections.

8.4 Agent-Based Model

Initial Conditions

n_0 sites are located according to the principle of maximal differentiation. m_0 consumers are uniformly distributed along the circle.

Market Entry and Exit

We assume that each site incurs an initial cost in order to enter the market. The goal of each site is to repay this initial debt to the bank as well as the interest. If after a given number of time steps the site is not able to fully repay its current debt to the bank, the site is forced to exit the market. If a site that belongs to a coalition dies the coalition's size is reduced by one.

Agents' Growth Rates and Localization

We assume that the number of consumers and the number of sites grow exponentially with time at rates g_c and g_s respectively. Therefore at time step t the number of consumers is $m(t) = (1 + g_c)^t m_0$ and the number of sites is $n(t) = (1 + g_s)^t n_0$. New sites and consumers are located as the initial populations.

Coalition Formation Process

Every T_c time steps single sites and/or coalesced sites are randomly activated to engage in a new coalition formation process. If the site is single, then it becomes a coalition initiator by randomly selecting a partner. If the selected partner is single as well, they form a new coalition of size 2. Otherwise, the initiator joins the existing coalition, thereby increasing its size by one. If the initiator already belongs to a coalition, then it randomly selects a single partner in order to expand its coalition by one member. In our model, we do not allow the merging of two existing coalitions. Once a new coalition is formed, it is tested in the market for the next T_c time steps. At the end of the test period, each member of the new coalition compares the sum of its individual profits over the last T_c time steps with the equivalent sum in the previous T_c time steps (whether they belonged to a coalition or not). If profits have increased for each site of the new coalition then we consider that the criterion of individual satisfaction is satisfied (the coalition is considered profitable for all participants) and the coalition is permanently adopted. Otherwise, the last member added in the coalition is not accepted. Sites that have been rejected by a given coalition will not be candidates for acceptance by the same coalition ever again in the course of the simulation.

Consumers' Search Process

In each time step consumers are randomly activated to undertake a new search process that may result in the purchase of one product unit, following the behavior described in the theoretical model.

RND and DFA Price Adjustment Procedure

In the RND algorithm (random algorithm), at each time step, each site readjusts its price after comparing the sum of profits that it had during the last T_P steps with the one of the previous T_P steps. If there has been an increase in profits, the site makes a price adjustment by increasing or decreasing the price by an amount selected randomly in $[\delta p_{min}, \delta p_{max}]$ (uniformly), otherwise the site readopts the price it had before the last price adjustment. In the DFA (Derivative Following Algorithm, see Kephart **et al.** (1998)), each site makes its first price adjustment randomly. If after T_P steps, the sites finds that its profits have increased, it keeps moving the price in the same direction, otherwise it reverses direction.

Consumers' Expertise and Reduction of Search Cost

Consumers improve with time their ability to search and find products on the Web, so we consider that the search cost of the consumers decreases linearly with time.

8.5 Simulations Results

8.5.1 Simulation plan

The first step of the simulation plan consists of keeping the symmetry assumptions that have been used in the analytical model in order to study the impact of the coalition formation procedure on the market structure. In this context, we analyze the role of the structural parameters of the model (*i.e.* adaptation cost and search cost) on the results. Next, we explore the effect of the market entry and exit processes (for sites and consumers) on the market structure. More precisely, we are interested in identifying those forces that are capable of creating *aggregation* in our simulated market. We define aggregation depending on the proportion of coalitions of each size. A market structure in with many big coalitions is considered to be *dispersed*, which results in an oligopolistic market. A market structure in which there is only a small number of big coalitions while the rest of the sites are either single or participate in small coalitions is considered to be *concentrated*.

8.5.2 Scenario 1

Reference simulation. We start by a generalization of the analytical model to 20 sites and where only the natural territories of consecutive sites intersect (in this case connexity corresponds to consecutive locations on the circle). There are 1000 homogeneous consumers (same search cost $c = 1$, same adaptation cost $t = 1000$ and reservation utility $R = 10000$, symmetric locations). The 20 sites are symmetric (same price, symmetric locations), and inside a coalition the search cost of the consumers is $c' = 0$. All sites are located according to the principle of maximum differentiation, whereas the consumers expect the distribution of sites' locations to be uniform. All sites set the same price $p^* = 10$, which is the price anticipated by consumers. The length of the corresponding natural territory is $2\sqrt{c/t} = 0.063$. We test two different coalition formation procedures: *(i)* a unique site is activated as initiator of a single coalition that grows throughout the simulation, and *(ii)* all sites can be activated and are able to form coalitions in parallel.

Results of Scenario 1

(i) In typical runs where only one coalition is allowed to grow, the single coalition formation lead to a coalition of size 5. *(ii)* Figure 8.2 shows a typical result of a simulation for the same parameters but allowing the formation of multiple coalitions with different initiators. For presentation reasons, instead of drawing all the coalitions on the same circle, coalitions are grouped by size.

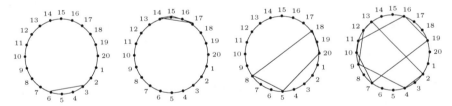

Fig. 8.2. A typical result for scenario 1

In this simulation, the multiple coalition formation led to the emergence of 5 coalitions (1 of size 2, 1 of size 3, 1 of size 4 and 2 of size 5). As we can see, the coalitions grew in general by adding a non-connex member each time. The only way to obtain a "connex component" (two members of the coalition are connex, like for the coalition of size 3) is that the coalition starts by two connex partners. This first result confirms the theoretical result in terms of coalition structure: *non-connex partners are preferred when forming a coalition.*

Random Effect and Coalitions' Size

In addition to this, we see that the maximum size of a coalition is 5, which is much smaller than what could be expected (a single coalition of size 10) by extending the result of the theoretical model (non-connex partners are preferable to connex ones). This is due to the fact that the number of consumers inside the natural territory of any site is very small, only $n.2\sqrt{c/t} = 1000 * 0.063 = 63$ consumers. As a result, the intersection of natural territories contains $1000 * 0.013 = 13$ consumers. Without coalitions, consecutive sites would share equally these 13 consumers. A coalition of size 2 with a non-connex site increases the probability to obtain these consumers from $1/2$ to $2/3$. This amounts to 8 consumers on expectation instead of 6. However, for larger coalitions, for example when the coalition size passes from size 5 to 6, the expected amount of consumers is $(6/7 - 5/6) * 13 = 1.53$. In other words, due to the combined effect of the number of consumers on the market (discrete distribution, instead of continuous distribution as in the theoretical model) and the length of natural territories, the advantage of being coalesced may not appear. This phenomenon will be referred to as *random effect*.

Competition Among Coalitions and Coalitions' Size

Added to this random effect, the marginal increase in profits to the last entrant in a coalition decreases with the size of the coalition. In a typical run, the average profit increased by 24% when the initiator found another single site to coalesce with, but as new sites were being added to the coalition the average profit increase became smaller (4%). In the case of multiple coalition formation (Figure 8.2), the maximum coalition size obtained was also 5. In the multiple coalition formation case, another phenomenon can have impact on the stopping of the growth of a coalition: the "competition" among existing coalitions for new highly differentiated members. As the coalitions grow in size, there are fewer highly differentiated sites that could be accepted by these coalitions. The importance of this result lies in the fact that a discrete distribution of consumer preferences is a more realistic setting than the continuum considered in the theoretical model. The outcome of the theoretical model according to which it is always profitable to coalesce has to be reconsidered since the way in which the number of consumers interacts with the values of the search cost and adaptation cost (that define natural territories) influences dramatically the size of the coalitions. As a result we can say that the coalition growth stops before it reaches the size of the maximum possible non-connex coalition.

8.5.3 Scenario 2 - Impact of Structural Parameters

The analysis of the impact of the parameters of the model reveals that only the values of search cost and adaptation cost have a major impact on the emerging coalition structures. Until now we studied the extreme case of natural territories intersecting only for consecutive neighbors. We present now the typical results from an experiment in which the values of the parameters are such that the natural territory of each site is so large that it almost covers the whole circle. All parameters are the same than in scenario 1, except for the adaptation cost $t = 5$ instead of 1000. The length of the sites' natural territory in this case is $2\sqrt{c/t} = 0.89$.

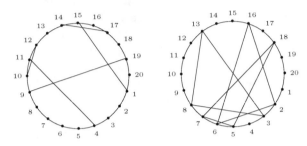

Fig. 8.3. A typical result for scenario 2

Results of Scenario 2

(i) In typical runs of simulations where only one coalition is allowed to grow, the process stopped at size 2. As we could expect, since there are less non-connex potential partners, the size of the coalition is smaller than in scenario 1. *(ii)* Figure 8.3 shows a typical result of a simulation allowing the formation of multiple coalitions with different initiators. We also see that only coalitions of small size emerged, consisting of 2-3 members (5 coalitions of size 2 and 3 coalitions of size 3). Given that there is extended overlapping among the natural territories in this experiment, one would expect that there would be no coalitions of size greater than 2, just like in the case of a single coalition, since only diametrically opposed sites have territories that do not intersect. Let us analyze the presence of "connex components" in some coalitions (the third site of a coalition being in this scenario always connex with the other two).

Existence of Connex Components

As we just mentioned, we observe that connex partners are accepted in coalitions. This phenomenon, which is impossible in the case of a single coalition, is made possible due to the dynamic process of multiple coalition formation. In order to explain it we analyze the chronology of the formation of the coalition (3,13,8) of Figure 8.3: sites 3 and 13, which are non-connex, formed a coalition of size 2. This allows site 3, for example, to attract more consumers on expectation over those that belong to the intersection of territories with connex sites that are not yet coalesced. Next, coalitions (2,16) and (5,18) formed. Site 3 is now surrounded by two connex neighbors that participate in coalitions of size 2. Therefore, the advantage of being coalesced for site 3 is diminished: the probability to obtain consumers belonging to the intersections of natural territories of sites 2, 3 and 5 is now identical for these three sites. Afterwards, site 8 enters coalition (3,13). Thus, we obtain a coalition of size 3 (3, 13, 8) with 2 connex components (13,8) and (3,8). Even though site 8 is connex with site 3, the participation of site 8 in the coalition is beneficial for site 3 which can now attract more consumers than sites 2 and 5. As a result we can say that *although non-connex partners are always preferable for coalescing with, the dynamic process of multiple coalition formation allows the emergence of coalitions containing connex components as a reply to the competition from existing coalitions in the market.*

8.5.4 Scenario 3 - Population Growth and Market Exit

We will now study scenarios where the populations of users and sites grow over time with exponential rate. There are 1000 consumers in the market, and this population will grow to 5000. The population of sites will grow from 20 to 100 sites. The values of the consumers' adaptation and search cost are such that the natural territories of each pair of consecutive sites that belong to the initial population of sites intersect. Concerning the locations of the new sites and users they follow the principal of maximal differentiation. The rest of the parameters and assumptions are the same as those of scenario 1.

Results of Scenario 3 (Constant Ratio of Consumers/Sites, No Market Entry/Exit)

In order to have a reference, we present first the results of an experiment where there is no exit process and the numbers of sites and users grow at exponential rate. In a typical run, 16 coalitions were formed: 1 of size 2, 2 of size 3, 2 of size 4, 4 of size 5, 6 of size 6 and 1 of size 7. Coalitions of final size from 5 to 7 were formed in the beginning of the simulation, whereas the majority of smaller size coalitions were formed in relatively late stages of the run. The first interesting phenomenon is that coalitions often surpass in size the maximum size found in the experiment of scenario 1 (no populations' growth). This is due to the fact that: (1) the growth of consumers reduces the random effect described earlier, thus allowing a coalition to grow more towards the size of the maximum non-connex coalition (size 10), and (2) the growth of sites changes the structure of the market in terms of natural territories, making that more sites are connex. Therefore, connex components are now likely to appear increasing the size of a coalition. As a result we can say that *the agents' growth allows the coalitions to grow more through the reduction of the random effect (consumers' growth) and the possibility to accept connex components (sites' growth).*

Results of Scenario 3 (Constant Ratio of Consumers/Sites, Market Entry/Exit)

Next, we show the results when sites incur an initial cost to enter the market and exit the market if they cannot repay it in a given period of time. The related parameters are the following: entry cost=3600, time to repay=8 time steps, interest rate=5%. In this experiment, 8 coalitions formed: 1 of size 2, 1 of size 3, 1 of size 4, 2 of size 5, 2 of size 6 and 1 of size 8. We can observe the coexistence of a high number of large size coalitions (5 coalitions of size more than 5). In this case, the market structure goes towards duopoly or monopoly structures.

Results of scenario 3 (Increasing Ratio of Consumers/Sites, Market Entry/Exit)

In this experiment 16 coalitions are formed (which is the double than what was obtained with constant consumers/sites ratio): 3 of size 2, 2 of size 3, 2 of size 4, 2 of size 5, 4 of size 6 and 3 of size 7. There is a significant number of big size coalitions (size 5-7). 7 of these coalitions have a size greater than 6. As a result we can say that *an increasing ratio of consumers to sites encourages the emergence of dispersion in the market resulting in an oligopoly of big coalitions.*

Results of Scenario 3 (Decreasing Ratio of Consumers/Sites, Market Entry/Exit)

In this experiment 22 coalitions are formed: 13 of size 2, 7 of size 3, 1 of size 5 and 1 of size 7. We observe aggregation in the emerging market structure: only 2 coalitions of size greater than 3 are formed, while there is a large number of small coalitions of size 2 and 3. We see that only 2 big coalitions dominate the market. The duopoly faces the competition of many small-size coalitions and single sites, which are less efficient. As a result we can say that *a decreasing ratio of consumers to sites encourages the emergence of concentration tending to a monopolistic market structure.*

Results of Scenario 3 (Decreasing Search Cost)

We suppose that the search cost decreases linearly for all consumers with time. The search cost at time step 1 is 1.0 for all consumers and 0.25 at the end of the simulation. The reduction of search cost has two implications.

On one hand the natural territories of sites, which are given by the formula $\sqrt{c/t}$, decrease with the search cost, meaning that with time sites become less connex among them (the intersection of their natural territories diminishes) and as a result, it is more easy for a site to accept to coalesced with another site. The simulations reveal that the same number of coalitions is formed as when the search cost is constant, but we do not observe the presence of a large size coalition. On the contrary, we observe more medium-sized coalitions. When the ratio of consumers to sites decreases with time (this setting lead to a concentrated market structure as mentioned before), we see that the market is less concentrated as the search cost decreases. This creates more opportunities for coalitions to be formed resulting in a market in which more coalitions manage to grow. As a result, the decrease in the consumers' search cost with time diminishes market concentration.

Results of Scenario 3 (Price Adjustment)

We suppose that sites have the possibility to adjust their prices according to the two alternative algorithms. We are interested in seeing whether or not the concentration that emerges in the market is affected. The sequence of events remain the same except for the fact that we suppose that sites adjust their prices in the beginning of each time step (between each period of search and consumption). The price adjustment is such that $\delta p_{min} = \delta p_{max} = 1.0$. The initial price of each site entering the market is randomly selected in $[0.1, 20.0]$. The experiments with the different price adjustment algorithms show that both kinds of price adjustment algorithms give better average profits for all sites, DFA being more efficient. However, we observe that the price adjustment is more beneficial for coalesced sites than for non-coalesced. In terms of aggregation, the DFA price adjustment algorithm diminishes less the concentrated character of the market compared to the RND price adjustment algorithm. As a result, *with price adjustment, the average profit of coalesced sites is better than the profit of non-coalesced sites, confirming the result of the static comparative model.*

8.5.5 Remark

All the experiments of the previous scenarios have been repeated with heterogeneity in the agents' characteristics (the corresponding values were selected following a uniform distribution): search cost, reduced search cost, reservation utility, adaptation cost, prices (constant) and entry cost. However, the determining value for these runs appeared to be the average value of these parameters resulting in similar results than those obtain with symmetric agents. The relative "noise" brought by variations in these parameters do not change the qualitative results.

8.6 Conclusion

This article summarizes the main results we obtained in an analytical model
and its agent-based extension for studying the emerging coalition structure in
electronic markets. The analytical model showed that the presence of search
costs for the consumers (that are independent from the adaptation costs) provides original and interesting results: *(i)* coalesced sites have an incentive to
lower their prices and *(ii)* sites choose a highly differentiated partner to form
a coalition. The agent-based model, enriched by additional behaviors for the
consumers and the sites, extends the analytical model with less restricting assumptions. The main findings of the agent-based model are the following: (1)
the dynamic process coalition formation confirms and generalizes the theoretical result in terms of coalition structure: non-connex partners are preferred
when forming a coalition, (2) the growth of the coalition is halted before it
reaches the size of the maximum possible non-connex coalition, (3) although
non-connex partners are always preferable for coalescing with, the dynamic
process of multiple coalition formation allows the emergence of coalitions containing connex components as a reply to the competition from existing coalitions in the market, (4) the agents' growth allows the coalitions to grow more
through the reduction of the random effect (consumers' growth) and the possibility to accept connex components (sites' growth), (5) an increasing ratio
of consumers to sites encourages the emergence of dispersion in the market
resulting to an oligopoly of big coalitions coexisting in the market, (6) a decreasing ratio of consumers to sites encourages the emergence of concentration
in the market tending to a monopolistic market structure, (7) the decrease
in the consumers search cost with time diminishes market concentration. Experiments performed on the role of heterogeneity proved to have little or no
impact on qualitative results mentioned above. If we look at the geographical
economy literature applied to the aggregation of shops in some locations, two
main forces are present: the lowering of consumers' search costs (whose main
component is transportation cost) and the increased competition due to the
proximity of shops. The first force drives aggregation by directly increasing
the demand. The second one may limit this aggregation due to an increase
in competition. In the world of B-to-C Internet sites, these mechanisms are
modified because search costs are independent from transportation costs and
because aggregation may arise through coalition of independent sites at (almost) no cost. Sites prefer to coalesce with differentiated partners, not for
avoiding stronger competition but for maximizing the increased market share
effect in the competition with non coalesced firms. The larger number of entry of new B-to-C Web-sites compared to the growth of consumers buying on
the Internet, excessively favored by the venture capitalism bubble for Web-
sites before the market was there, seems to be the main factor explaining the
market concentration on Internet.

References

[1] Axtell R (2000) Why agents? On the varied motivations for agent computing in the social sciences. The Brookings Institution, Working Paper

[2] Bakos Y (1997) Reducing buyer search costs: implications for electronic market places. Management Science 43:12:1676–1692

[3] Gabszewicz JJ, Garella P (1986) 'Subjective' price search and price competition. International Journal of Industrial Organization 4:305–316

[4] d'Inverno M, Kinny D, Luck M, Wooldridge M (1997) A Formal Specification of DMARS. In: Intelligent Agents IV: Proceedings of the Fourth International Workshop on Agent Theories, Architectures, and Languages, ATAL-97. Springer LNAI 1365

[5] Kephart JO, Hanson JE, Levine DW, Grosof BN, Jakka Sairamesh, Segal RB, White SR (1998) Dynamics of an Information-Filtering Economy. In: proceedings of the Second International Workshop on Cooperative Information Agents (CIA-98)

[6] Kinny D (1999) A Framework for Multi-Agent Systems Development. In: Proceedings of the First Asia-Pacific Conference on Intelligent Agent Technology, Hong Kong

[7] Kinny D (2001) Reliable Agent Communication - A Pragmatic Perspective. New Generation Computing 19:2

[8] Laye J (2003) Price Competition And Coalition Strategy In The Presence Of Search Costs. PhD Thesis, Chapter 1, Ecole Polytechnique, Paris, France

[9] Laye J, Tanguy H (2005) Are the Neighbors Welcome? E-buyer Search, price competition and Coalition Strategy in the Internet Retailing. In: Brousseau E, Curien N (eds.) Internet and Digital Economics. Cambridge University Press

[10] Lina C (2003) Web formation model. In: I-Cities project, IST, 11337, Information Cities

[11] Wooldridge M, Jennings NR, Kinny D (2000) The Gaia Methodology for Agent-oriented Analysis and Design. Autonomous Agents and Multi-Agent Systems 3.

9

Agent Based Modeling of Trust Between Firms in Markets

Alexander Gorobets[1] and Bart Nooteboom[2]

[1] Sevastopol National Technical University, Management Department,
Streletskaya Bay, Sevastopol 99053, Ukraine `alex-gorobets@mail.ru`
[2] Tilburg University, P.O. Box 90153,NL-5000 LE Tilburg, the Netherlands
`B.Nooteboom@uvt.nl`

Summary. In this paper the methodology of Agent-Based Computational Economics (ACE) is used to explore under what conditions trust is viable in markets. The emergence and breakdown of trust is modeled in a context of multiple buyers and suppliers. Agents develop trust in a partner as a function of observed loyalty. They select partners on the basis of their trust in the partner and potential profit. On the basis of realized profits, they adapt the weight they attach to trust relative to profitability, and their own trustworthiness, modeled as a threshold of defection. Trust turns out to be viable under fairly general conditions.

9.1 Introduction

The viability of trust between firms in markets is a much-debated issue [9]. Economics, in particular transaction cost economics (TCE), doubts the viability of trust, on the argument that under competition, in markets, firms behave opportunistically in favour of profit [13]. Thus, disproving skepticism from TCE, it is of some theoretical and practical importance to explore under what conditions trust can be viable. TCE proposes that people organize to reduce transaction costs, depending on conditions of uncertainty and specific investments, which yield switching costs and a resulting risk of hold-up. In this paper we employ TCE logic, but we also deviate from TCE in two fundamental respects.

First, while TCE assumes that optimal forms of organization will arise, yielding maximum efficiency, we consider that problematic. The making and breaking of relations between multiple agents with adaptive knowledge and preferences may yield complexities and path-dependencies that preclude the achievement of maximum efficiency. Second, while TCE assumes that reliable knowledge about loyalty or trustworthiness is impossible [12, 13], so that opportunism must be assumed, we expect that under some conditions trust is feasible, by inference from observed behaviour, and that trustworthiness is viable, in yielding profit. To investigate this, the methodology of ACE enables us to take a process approach to trust [4, 15, 14], by modeling the adaptation of trust and trustworthiness in the light of experience in interaction.

The analysis is conducted in the context of transaction relations between multiple buyers and suppliers, where buyers have the option to make rather than buy, which is the classical setting for the analysis of transaction costs. We employ a model developed from an earlier model from Klos and Nooteboom [7]. In this model, agents make and break transaction relations on the basis of preferences, based on trust and potential profit.

The paper proceeds as follows. First, further specification is given of technical details of the model, needed to fully understand the experiments. Next, we specify the experiments and present the results. The paper closes with conclusions.

9.2 The Model

9.2.1 Preference and Matching

In the literature on trust distinctions are made between different kinds of trust, particularly between competence trust and intentional trust [9]. Intentional trust refers, in particular, to presumed absence of opportunism. That is the focus of TCE and also of the present paper. We focus on the risk that a partner will defect and thereby cause switching costs. In our model trust may be interpreted as a subjective probability that expectations will be fulfilled [2], which here entails realization of potential profit. Thus, expected profit (E) would be: E = profitability·trust. However, in the model, agents are allowed to attach more or less weight to trust relative to potential profit (α), on the basis of a generalized preference score:

$$\text{score}_{ij} = \text{profitability}_{ij}^{\alpha_i} \cdot \text{trust}_{ij}^{1-\alpha_i} \qquad (9.1)$$

where: score_{ij} is the score i assigns to j, $\text{profitability}_{ij}$ is the profit i can potentially make 'through' j, trust_{ij} is i's trust in j and $\alpha_i \in [0, 1]$ is the weight i attaches to profitability relative to trust, i.e. the 'profitelasticity' of the score. α is adaptive, as a function of realized profit. This 'Cobb-Douglas' function adopted from the literature on production functions, entails that profitability and trust are complements (they both contribute to the preference score) as well as substitutes (less profitability can be compensated with more trust).

At each time step, all buyers and suppliers establish a strict preference ranking over all their alternatives. Random draws are used to settle the ranking of alternatives with equal scores. The matching of partners is modeled as follows. On the basis of preferences buyers are assigned to suppliers or to themselves, respectively. When a buyer is assigned to himself this means that he makes rather than buys. In addition to a preference ranking, each agent has a 'minimum tolerance level' that determines which partners are acceptable. Each agent also has a quota for a maximum number of matches it can be involved in at any one time. A buyer's minimum acceptance level of suppliers is the score that the buyer would attach to himself. Since it is reasonable that he completely trusts himself, trust is set at its maximum of 1, and the role of trust in the score is ignored: $\alpha = 1$. The algorithm used for matching is a modification of Tesfatsion's [11] deferred choice and refusal (DCR) algorithm and it proceeds in a finite number of steps, as follows:

1. Each buyer sends a maximum of o_i requests to its most preferred, acceptable suppliers. Because the buyers typically have different preference rankings, the various suppliers will receive different numbers of requests.
2. Each supplier 'provisionally accepts' a maximum of a_j requests from its most preferred buyers and rejects the rest (if any).
3. Each buyer that was rejected in any step fills its quota o_i in the next step by sending requests to next most preferred, acceptable suppliers that it has not yet sent a request to.
4. Each supplier again provisionally accepts the requests from up to a maximum of a_j most preferred buyers from among newly received and previously provisionally accepted requests and rejects the rest. As long as one or more buyers have been rejected, the algorithm goes back to step 3.

The algorithm stops if no buyer sends a request that is rejected. All provisionally accepted requests are then definitely accepted.

9.2.2 Trust and Trustworthiness

Trust, taken as inferred absence of opportunism, is modelled as observed loyalty, i.e. observed absence of defection. Following Gulati [4], we assume that trust increases with the duration of a relation. As a relation lasts longer, i.e. there is no defection, one starts to take the partner's behaviour for granted, and to assume the same behaviour (i.e. commitment, rather than breaking the relation) for the future. Thus, agent i's trust in another agent j depends on what that trust was at the start of their current relationship and on the past duration of that relationship:

$$t_i^j = t_{init,i}^j + (1 - t_{init,i}^j)\left(1 - \frac{1}{fx + 1 - f}\right) \tag{9.2}$$

where

t_i^j = agent i's trust in agent j,
$t_{init,i}^j$ = agent i's initial trust in agent j,
x = the past duration of the current relation between agents i and j, and
f = trustFactor.

This function is taken simply because it yields a curve that increases with decreasing returns, as a function of duration x, with 100% trust as the limit, and the speed of increase determined by the parameter f.

In addition, there is a base level of trust, which reflects an institutional feature of a society. It may be associated with the expected proportion of non-opportunistic people, or as some standard of elementary loyalty that is assumed to prevail. If an agent j, involved in a relation with an agent i, breaks their relation, then this is interpreted as opportunistic behaviour and i's trust in j decreases; in effect, i's trust drops by a percentage of the distance between the current level and the base level of trust; it stays there as i's new initial trust in j, $t^j_{\text{init},i}$ until the next time i and j are matched, after which it starts to increase again for as long as the relation lasts without interruption.

The other side of the coin is, of course, one's own trustworthiness. This is modelled as a threshold τ for defection. One defects only if the advantage over one's current partner exceeds that threshold. It reflects that trustworthiness has its limits, and that trust should recognize this and not become blind [10, 9]. The threshold is adaptive, as a function of realized profit.

9.2.3 Costs and Profits

Profit has the following elements. First, buyers may increase returns by selling more differentiated products. Second, suppliers may reduce costs by generating production efficiencies. There are two sources of production efficiency: economy of scale from a supplier producing for multiple buyers, and learning by cooperation in ongoing buyer-supplier relations. Economy of scale can be reaped only in production of standardized products, with general-purpose assets, and learning by cooperation can only de achieved in production that is specific for a given buyer, with buyer-specific assets.

This yields a link with the fundamental concept, in TCE, of 'transaction specific investments'. We assume a connection between the differentiation of a buyer's product and the specificity of the assets required to produce it. In fact, we assume that the percentage of specific products is equal to the percentage of specific assets. This is expressed in a variable $d_i \in [0, 1]$. It determines both the profit the buyer will make when selling his products and the degree to which assets are specific, which determines opportunities for economy of scale and learning by cooperation. This parameter is part of the 'state of the world', in this case the market, and applies to all agents, in a given run of the model.

Economy of scale is achieved when a supplier produces for multiple buyers. To the extent that assets are specific, for differentiated products, they cannot be used for production for other buyers. To the extent that products are general purpose, i.e. production is not differentiated, assets can be switched to produce for other buyers. In sum, economy of scale, in production for multiple buyers, can only be achieved for the non-differentiated, non-specific part of production, and economy by learning by cooperation can only be achieved for the other, specific part.

Both the scale and learning effects are modelled as follows:

$$y = \max\left(0, 1 - \frac{1}{fx + 1 - f}\right) \tag{9.3}$$

where:

For the scale effect $f=$ scaleFactor, x is general-purpose assets of supplier j summed over all his buyers. Here, y denotes scale efficiency achieved by supplier j.

For the learning effect $f=$ learnFactor; x is the number of consecutive matches between supplier j and buyer i. Here, y denotes learning efficiency achieved in the collaboration between supplier j and buyer i.

Formula (9.3) expresses decreasing returns for both scale and experience effects. The scale effect is specified in such a way that a supplier can be more scaleefficient than a buyer producing for himself only if the scale at which he produces is larger than the maximum scale at which a buyer might produce for himself. For the learning effect, a supplier's buyerspecific efficiency is 0 in their first transaction, and only starts to increase if the number of transactions is larger than 1. If a relation breaks, the supplier's efficiency due to his experience with the buyer drops to zero.

All this results in the following specification of profit. The number of generalpurpose assets that a supplier j needs in order to produce for a buyer i, is equal to $(1 - d_i)(1 - e_{s,j})$, where $e_{s,j}$ is scale efficiency $(0< \quad e_{s,j} \quad <1)$ in the production volume of supplier j. The number of buyer-specific assets that a supplier j needs, to produce for a buyer i, is equal to $d_i(1 - e^i_{l,j})$, where $e^i_{l,j}$ is learning efficiency $(0< \quad e^i_{l,j} <1)$ in the relationship between buyer i and supplier j. Thus, the profit that can potentially be made in a transaction between a buyer i and a supplier j is:

$$p^j_i + p^i_j = (1 + d_i) - (d_i(1 - e^i_{l,j}) + (1 - d_i)(1 - e_{s,j})). \tag{9.4}$$

The first part of the formula specifies returns and the second part specifies costs. It is assumed that the agents involved share the profit equally.

9.2.4 Adaptation

An agent is adaptive if 'the actions of the agent in its environment can be assigned a value (performance, utility, payoff, or the like); and the agent behaves in such a way as to improve this value over time' [5]. In this model, agents adapt the values for $\alpha \in [0, 1]$ (weight attached to profit relative to trust) and τ [0, 0.5] (threshold of defection) from one time step to the next, which may lead to changes in the scores they assign to different agents. The idea is that the values that have led to high performance (profit) increase probability that those values will be selected again. This is a simple model of 'reinforcement learning' [1, 6].

While τ could conceivably rise up to 1, a maximum of 0.5 was set because initial simulations showed that otherwise relations would get locked into initial situations, with little switching. Note that this biases the model in favour of opportunism.

9.2.5 The Algorithm

The algorithm of the simulation is presented by the flowchart in Figure 9.1. This figure shows how the main loop is executed in a sequence of discrete time steps, called a 'run'. Each simulation may be repeated several times as multiple runs, to even out the influence of random draws in the adaptation process. At the beginning of a simulation, starting values are set for certain model parameters. The user is prompted to supply the number of buyers and suppliers, as well as the number of runs, and the number of timesteps in each run.

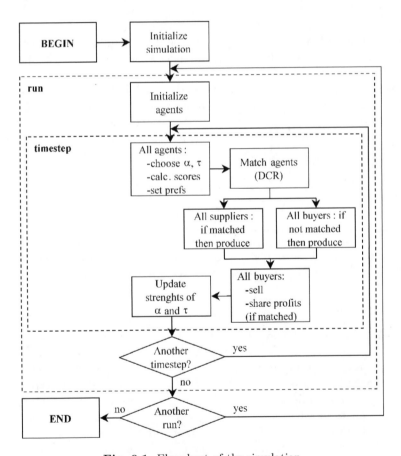

Fig. 9.1. Flowchart of the simulation

9.3 Experiments

9.3.1 Hypotheses

The goal of the experiments is to test the following hypotheses.

Counter to TCE we expect:

Hypothesis 1 *Due to complexities of interaction maximum efficiency can rarely be attained.*

Hypothesis 2 *Even in markets, where profit guides adaptation, high trust(low α; high τ) may be sustainable.*

Hypothesis 3 *The choice between an opportunistic switching strategy and loyalty depends on the relative strength of scale effects and learning by cooperation*

In agreement with TCE we expect:

Hypothesis 4 *When trust is low, higher asset specificity/differentiated products yields less outsourcing.*

Hypothesis 5 *The more trust, the more collaboration in 'buy', rather than 'make'.*

Recall that if during the matching between buyers and suppliers a buyer decides to 'buy' rather than 'make', he can follow two different strategies. One is an opportunistic scale strategy, where the buyer seeks a profit increase on the basis of economy of scale, by trying to find a supplier who serves more than one buyer. This entails much switching and less emphasis on loyalty and trust. The other strategy is the learning by cooperation strategy, seeking an increase of profit in ongoing relations. This entails less switching and more emphasis on loyalty and trust. Thus, in manipulating the strength of the scale effect relative to the effect of learning by cooperation, we can bias the model towards opportunism or loyalty. This interacts with the degree of asset specificity/specialization, since economy of scale applies only to general purpose assets, and learning by cooperation only to specific assets. Note that there is an overall bias towards the opportunistic scale strategy, in that economy of scale is immediate, thus yielding a more immediate return in profits, while learning by cooperation takes time to build up. Thus, we are stacking the odds in favour of the TCE theory that we criticize. However, this does seem to be a realistic feature, supporting the intuition that trust is more viable in a long-term perspective.

9.3.2 Model Parameters

Each simulation run involves 12 buyers and 12 suppliers and continues for 100 timesteps. In order to reduce the influence of random draws, each run is repeated 25 times and results are averaged across all runs. Initially, results are also averaged for the two classes of agents: buyers and suppliers, in order to explore systematic effects. Each buyer's offer quota (maximum number of suppliers used) was fixed at 1, and each supplier's acceptance quota (maximum number of customers) was set to 3. In previous experiments with each supplier j's acceptance quota set to the total number of buyers, the system quickly settled in a state where all buyers buy from a single supplier. For this reason, suppliers were only allowed to have a maximum of three buyers. This limits the extent of the scale economies that suppliers can reach. A maximum number of buyers may be associated with competition policy setting a maximum to any supplier's market share.

For the test of our hypothesis, we consider different values for the percentage of specific assets/differentiated products: $d=$ 25, 45, and 65 %. We vary initial trust in the range 10, 50 and 90%, initial threshold for defection (τ) from 0 to 0.5, initial weight attached to profit relative to trust (α) from 0.0 to 1.0, and the fixed parameters of both the strength of economy of scale and learning by cooperation from 0.5 to 0.9.

9.4 Results

We present the results in the order of different starting values of trust. This reflects different institutional settings, from high to low trust 'societies'. Here, we can see to what extent those are stable or shift. In particular, the question is whether high initial trust can be sustained, and whether perhaps distrust can evolve into trust.

9.4.1 An Individual Trajectory

First to give some feel of what is going on, we present and discuss time paths for selected individual relations, to illustrate the process of switching, or lack of it. Subsequently, we present the overall results, in comparison with our expectations, on the basis of averages, across runs as well as agents (all buyers, all suppliers).

Since we are particularly interested in what happens at high trust, we select high trust case from a simulation with parameters: Initial Trust is 90%, the weight attached to profit relative to trust (α) is zero, the threshold of own defection (τ) is at its maximum of 0.5, the factors indicating the strength of learning and economy of scale (Learnfactor & Scalefactor) had an intermediate value of 0.5. In this case, which is biased towards trust and loyalty, agents are expected to favour loyalty and the learning by cooperation strategy.

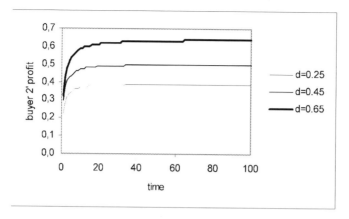

Fig. 9.2. Buyer 2's profit

Figure 9.2 shows the actual profit for buyer 2. For the run with d=0.25 buyer 2 buys from supplier 6, for d=0.45 from supplier 3 and for d=0.65 from supplier 2. In his relation with supplier 6, for $d = 0.25$, his trust in that supplier increases up to almost the maximum of 1.0 (see Fig. 9.3). Not shown is the result that the same applies to supplier's trust in buyer 2. For other values of d buyer 2 has no relation with supplier 6, so that his trust does not get any opportunity to rise. Supplier 6's profit is shown in Figure 9.4. For $d = 0.25$, he offers a big scale effect, producing for 3 buyers simultaneously: buyers 4 and 6 in addition to buyer 2. For $d = 0.65$, supplier 6 has 2 buyers, nr. 5 and 6, and he doesn't have any buyer for d=0.45. Not shown is the result that supplier 6's weight attached to profit relative to trust (α) and threshold of own defection (τ) remain about the same as their starting values. As shown in Figure 9.5, buyer 2 at first thinks he can increase profit by increasing the weight attached to profit relative to trust (α), but then learns that this does not work and reduces it again. Not shown is that his threshold of own defection (τ) remains at its initial value of 0.5.

9.4.2 Overall Results

Now we turn to the more representative, overall results, in terms of averages across agents and runs (for the details see Gorobets and Nooteboom [3]).

All our expectations are borne out by the experiments. Of course, simulation is not equivalent to empirical testing. The test is virtual rather than real. We have only shown that under certain assumptions emergent properties of interaction satisfy the hypothesis. The significance of this depends on how reasonable the assumptions in the model are considered to be.

If we compare across the different settings of high, medium and low trust, under the same conditions concerning the relative strength of scale effect, learning by cooperation, and under the same initial conditions concerning

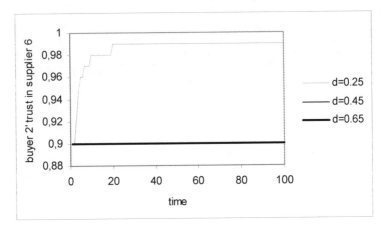

Fig. 9.3. Buyer 2's trust in supplier 6

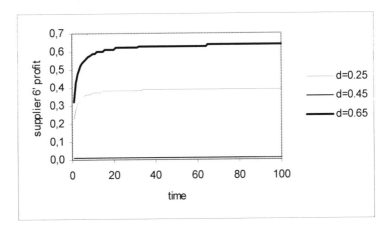

Fig. 9.4. Supplier 6's profit

weight attached to trust and threshold of defection, profit declines more often than it increases, as we go from high to low trust.

Overall, the results can be summarized as follows. A strong effect of learning by cooperation, a high weight attached to trust, and high loyalty favour the learning by cooperation strategy for high levels of specific investments, while a high weight attached to profit and high loyalty favour the scale strategy for low and average levels of specific investments.

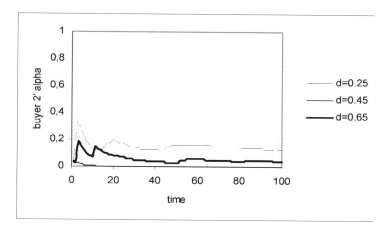

Fig. 9.5. Buyer 2's α

9.5 Conclusions

The general outcome is that in interactions between agents both trust and opportunism can be profitable, but they go for different strategies. This suggests that there may be different societies, going for different strategies, of switching or of loyalty, which settle down in their own self-sustaining systems.

References

[1] Arthur, W. Brian.(1993) On designing economic agents that behave like human agents.Journal of Evolutionary Economics, 3/1 1-22.
[2] Gambetta, D. (ed.)(1988) Trust: The Making and Breaking of Cooperative Relations. Basil Blackwell, Oxford
[3] Gorobets, A., Nooteboom, B. (2005) Agent Based Computational Model of Trust. In: Czap, H. et al.(eds.)Self-Organization and Autonomic Informatics (I), Vol. 135. *IOS*Press, Amsterdam 160-172
[4] Gulati, R.(1995) Does familiarity breed trust? The implications of repeated ties for contractual choice in alliances. Academy of Management Journal,38/1 85-112
[5] Holland, J.H., Miller, J.H.(1991) Artificial adaptive agents in economic theory. American Economic Review, 81/2 365-370.
[6] Kirman, A.P., Vriend, N.J.(2001) Evolving market structure: An ACE model of price dispersion and loyalty. Journal of Economic Dynamics and Control, 25/3 (2001) 459-502
[7] Klos, T.B., Nooteboom, B.(2001) Agent based computational transaction cost economics. Journal of Economic Dynamics and Control, 25 503-526.

[8] Lewicki, R.J., Bunker, B.B.(1996) Developing and maintaining trust in work relationships. In: Kramer, R.M., Tyler, T.R. (eds.) Trust in organizations: Frontiers of theory and research. Thousand Oaks, Sage Publications 114-139

[9] Nooteboom, B.(2002) Trust: forms, foundations, functions, failures and figures. Edward Elgar, Cheltenham UK

[10] Pettit, Ph.(1995) The virtual reality of homo economicus. The Monist, 78/3 308-329

[11] Tesfatsion, L.S.(1997) A trade network game with endogenous partner selection'. In: Amman, H.M., Rustem, B., Whinston, A.B. (eds.) Computational Approaches to Economic Problems. Advances in Computational Economics, Vol. 6. Kluwer, Dordrecht 249-269

[12] Williamson, O.E.(1975) Markets and Hierarchies: Analysis and Antitrust Implications. The Free Press, New York

[13] Williamson, O.E.(1993) Calculativeness, trust, and economic organization. Journal of Law and Economics, 36/1 453-486

[14] Zand, D.E.(1972) Trust and managerial problem solving. Administrative Science Quarterly, 17/2 227-239.

[15] Zucker, L.G.(1986) Production of trust: Institutional sources of economic structure, 1840- 1920. In: Staw, B.A., Cummings, L.L. (eds.) Research in Organizational Behavior, Vol. 8. JAI Press, Greenwich Conn 53-111

10

Investigations into Schumpeterian Economic Behaviour Using Swarm

Craig Lynch

Macquarie University, Sydney, Australia, `clynch@tpg.com.au`

10.1 Introduction and Background

Ever since the field of economics emerged there has been continual debate over the mechanisms and drivers of economic growth. Two conceptual views that are diametrically opposite are embodied in the fields of macroeconomics and microeconomics. Both of these arenas have different philosophical foundations, macroeconomics being predominantly focused on the behaviour of complete economies reacting to external stimuli, whilst microeconomics is concentrated on the behaviour of the smallest components of an economy, namely the various actors contained within it, interacting to form an aggregate of economic behaviour.

The macroeconomic view sees the processes of economic behaviour as being relatively static, with the belief that these processes inherently lead to an equilibrium state. A change in the environment the economy operates within will alter this equilibrium, once the change has concluded, the economy will return to another equilibrium position. Specific relationships between behaviour and outcomes at a granular level, for example the link between entrepreneurship and wealth creation, are ignored.

The microeconomic view, in contrast, places the greatest importance on individuals and their decision making, the core belief being that human behaviour isn't necessarily rational and processes inherent in the working economy fluctuate depending on individual or firm behaviour, and can't consequently be considered efficient or even necessarily predictable.

Whilst there are many areas of overlap between these perspectives, there are noticeable gaps, specifically in the areas of entrepreneurial impact, the role of innovation, the process of entry and exit of firms in the marketplace, and the relative efficiency within firms of introducing innovation.

The microeconomic view is drawn from Schumpeter's Theory of Economic Development, the underlying focus being on individual human behaviour which, when aggregated, represents building blocks for the overall economy, an approach that can be seen as akin to social science (Shionoya 1997). One of the theory's focal areas is the perspective that economic growth is driven by dynamic behaviour amongst all economic actors, and equilibrium is never reached – the competitive behaviour inherent in the marketplace continually creates disequilibrium.

Central to the theory is the role of the entrepreneur, the changing cycle of innovation, and the dynamic interplay between firms operating in a highly competitive environment. Schumpeter places the entrepreneur as the central figure causing disequilibrium (Clemence and Doody 1966) whilst other perspectives see them taking advantage of disequilibrium conditions rather than being a cause (Kirzner 1973). Within this context, two areas of economic postulation and analysis have emerged over the past several decades - *evolutionary economics* and *industrial dynamics*. Whilst not being mutually exclusive, these new areas view economic growth from different perspectives, attempting to understand the intrinsic reasons for growth using different concepts and methods of analysis.

The evolutionary economics concept holds that the economy can be perceived as a large population of diverse agents, incorporating technology change and self-transformation of industry structures (Nelson 1994). From this, empirical studies with this population perspective are feasible and interactive behaviour within the population can theoretically be modeled. With the availability of modern computational tools, much of the focus in this area has moved toward formal simulation modeling, where multiple combinations of diverse economic elements can be synthesized and overall economic growth examined under varying environmental conditions.

In comparison, industrial dynamics represents a desire to more fully understand the intrinsic behaviour between all participants in an economic environment, whilst still attempting to understand the underlying reasons for economic growth at a higher and observable level. As quoted, "Economic growth can be *described* at the macro level, but it can never be *explained* at that level" (Carlsson and Eliasson 2001). The broad environment is still understood to be characterised by essentially Schumpeterian behaviour, however the ideal focus within industrial dynamics is to quantify this behaviour within the more general purpose of integrating the specific behaviour by specific actors (whether they are firms, consumers, institutions, or entrepreneurs) to the theoretical and empirically observable behaviour inherent in a dynamic environment.

10.2 Key Issues and Objectives

Taking into account the background provided to this point, and recognising the ongoing debate within the economic and industrial dynamics arenas concerning the fundamental nature of economic growth and how a competitive inter-firm environment fundamentally operates, there is clearly a need to explore the attributes of this environment in greater detail and to focus on the impacts introduced by diverse competitive behaviours.

From a system modeling perspective, there is a fundamental question at hand – can the Schumpeterian environment representing a complex economic system be adequately modeled to allow greater insight into these issues? In particular, can a model be established that can operate both at a holistic, aggregated level and a lower level where the specific behaviour of individual actors in the economy can be isolated and quantified? For this second level, can industrial dynamics be explored to as granular level as possible?

The focus of this paper is to highlight the relationships between macroeconomic effects and the diverse microeconomic actions implicit in firm decision-making, so as to understand their aggregative effects on an economy.

The methodology undertaken was to establish a model suitable for a range of Schumpeterian behaviours to operate within, and be able to isolate the impact of specific competitive behaviour for subsequent analysis and exploration. The assumption is that economic growth is endogenous and an outcome of interactions at a micro level rather than being the outcome of external factors (Metcalfe 1998).

Within this context, the sub-objectives of the project were to explore the following specific economic and competitive behavioural concepts:

1. The impact of random innovation on firms within an economy.
2. The availability of credit and differentiation of supply to entrepreneurs.
3. The impact of predator behaviour amongst firms.
4. The impact of innovation investments on firm growth.
5. The impact of imitation and marketing defense investments on firm growth and survival.

10.3 Model Design

The model used for this project was adapted from the agent-based simulation model developed by Bruun and Luna in 2000 to examine endogenous economic growth within an artificial economy. This model established a foundation environment where the interaction between consumers, workers/artisans, entrepreneurs and firms defined the economy. Activity in the model was in the form of consumer purchasing behaviour within a context of firm creation and dissolution. The dynamic nature of agents adapting to different roles and the incorporation of behavioural randomness within run-time parameters was a key feature of the model, ensuring that as much as possible it represented a "real" economy where individual decision-making cannot be predicted.

For this project, the Bruun and Luna model has been enhanced to allow a specific focus on five key environmental conditions, namely (1) how the economy behaves at a macro level to the random introduction of innovation at a firm level, and how this influences behaviour at an agent type level; (2)

how behaviour is influenced when a relationship between credit availability
and proven entrepreneurial competencies is introduced; (3) how firms grow
or change when predator behaviour is allowed, represented by the acquisition
of one firm by another; (4) how growth is influenced by the potential for
innovation to be established by strategic decision-making within a firm to
allocate part of its profits into research and development expenditure; and (5)
the relatively more complex scenario where firms can imitate other firms at a
product level, using a proportion of profits to do so, and with the consequent
ability of firms to also establish a marketing-type defense against imitation
using proportions of profits.

Run-time execution and subsequent analysis is based on the establishment
of six discrete stages within the model. Each stage can be activated or deac-
tivated as a starting parameter depending the analysis purpose of each run.
The first stage represents the core Bruun and Luna model and is always activated, being the foundation for all other stages to operate. In some cases,
stages must be activated in conjunction with others given the key variables
introduced into the model by these other stages.

Three new agent-level variables have been incorporated to support these
scenarios. A primary attribute for firms is Product Attractiveness Level, representing a key change from the original model in consumer purchasing decisions
whereby preference is given to firms with products at a higher attractiveness
level. The mechanisms for changing individual attribute levels vary within the
different stages of the model – it could be derived directly from random innovation, an outcome from research and development expenditure, or the result
of firm imitation or marketing defense decisions. In all cases, the determination of the changed level is influenced by the degree of randomness inherent
within each stage.

Another important new attribute for firms is Financial Stability Level.
This is calculated and changed in a deterministic manner, representing the
degree of successful sales made by a firm over time, and increasing each time
a quantum number of sales are achieved. In a similar fashion to Product Attractiveness Level, the stability level of individual firms is used in a range of
agent behavioural decisions in multiple stages. Firm acquisition and imitation
and defense decisions are directly influenced by this attribute, and the outcomes of each firm decision also have a corresponding impact on the value of
the stability level. To some extent this level is analogous to profit, however in
a simplistic way. All decisions to act based on Financial Stability Level are
influenced by randomness parameters.

Given the focus on entrepreneurs at both theoretical and model design
levels, an Entrepreneur Competency level attribute has also been established.
This is primarily used in Stage 2, where the risk undertaken by lending in-
stitutes in providing credit to entrepreneurs is examined under conditions of
different credit being available to different entrepreneurs depending on the
competency level of each entrepreneur changing over time. Credit availability
is critical to Schumpeter's theory and is seen as crucial to entrepreneurial

operation (Bloch 2000), whilst the risk borne by lenders to entrepreneurs is based on opinion rather than firm knowledge (Knight). Randomness principles also apply to credit availability decisions.

One of the design features for all stages is the degree of randomness applicable to almost all new variable settings at the run-time level. Whilst the inclusion of randomness is in line with the core concepts of agent based modeling and represents the unpredictability of human behaviour, there are options within the model to increase or decrease the probability of random events from occurring. This capability, already available for some variables in the Bruun and Luna model, ensured that experimental runs were carried out in a relatively controlled but non-deterministic manner.

10.4 Results

10.4.1 Stage 1 – Random Innovation Impact Analysis

The design intent behind this stage of the model was to examine the impact of introducing random product innovation on both macro and micro aspects of the economy. This form of innovation is based on the early Schumpeterian concept of innovation, where innovation is completely random and cannot be consciously planned and introduced. The baseline version of the model works on the principle that all firms supply a ubiquitous product, with consumers changing preferences, or moving their shopping carts, through random settings within the model's environment.

The new firm attribute introduced in this stage, product attractiveness level, provides a foundation for consumers to evaluate the offering of a firm, and to make choices based on a scan of all firms in the immediate vicinity of their shopping cart. Increasing the product attractiveness attribute of firms is random, the degree of which being influenced by the random innovation seed. The lower the setting of this seed the greater the probability of random product attractiveness level increases within firms.

The model was run with five scenario settings – very high randomness to very low. A very low scenario (representing negligible, or zero, innovation) was deliberately sought to ensure that an equivalent baseline model execution could be achieved, and hence validate the underlying logic of the original model.

The results were interpreted using both macro and micro methodologies. Graphs of GDP over time for all aggregate firms provide insight into the overall performance of the economy, whilst individual firm and consumer decisions at each time period were extracted and analysed so as to focus on numbers and durations of effective firms, net account levels per firm, and hence calculate the persistence of profit for effective firms, this being seen as a factor of unique firm and industry attributes (Jacobson and Hanson 2001).

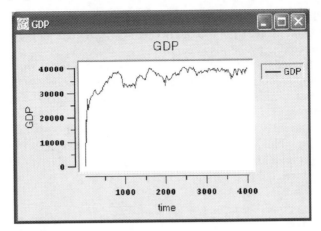

Fig. 10.1. LOW Innovation Randomness Seed

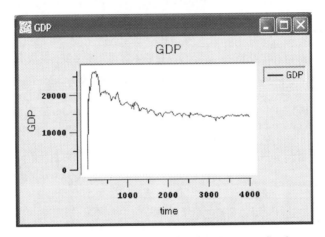

Fig. 10.2. HIGH Innovation Randomness Seed

Figures 10.1 and 10.2 represent two extreme GDP profiles out of all scenarios reviewed. From this, a small amount of innovation resulted in the highest GDP peak in the earliest time frame whilst any greater innovation caused the economy to degrade significantly.

The analysis outlined in Table 10.1 provides a more in-depth perspective at the firm level. As to be expected, the baseline (no innovation) scenario had the highest number of effective firms for the longest duration, and whilst a low innovation seed resulted in the highest GDP it was not favorable to individual firms. As innovation increased, profit increased although numbers of firms decreased.

Table 10.1. Random Innovation Scenarios

Seed Settings	No. Effective Firms	Average Duration of Firm (Time Periods)	Average Net Account Level per Firm	Persistence of Profit (Av Net Account Level / Firm Duration)
100 (High)	81	778	762,522	270
1,000 (Med – High)	99	965	433,385	69
10,000 (Medium)	119	1,465	44,340	12
100,000 (Low)	122	1,420	25,323	2
100,000,000 (V. Low)	123	1,471	91,102	53

10.4.2 Stage 2 – Risk / Entrepreneur Competency Evaluation

As part of the framework for credit supply decisions to be made, two agent attributes were used in this stage – financial stability level and entrepreneur competency level. An entrepreneur's financial stability level increases incrementally every time 100 demand signals have been successfully met. The entrepreneur's competency level increases incrementally when the combination of a successful credit request has been achieved and followed by the increase of financial stability level. The outcome of these two levels in place is the increased likelihood that an entrepreneur with a higher competence level and hence a better history of success in utilising investments will be given preference by banks in comparison to those with less success. Within this stage, with more credit the entrepreneurs in question will be in a better position to attract workers.

The model environment is influenced by two variables, credit availability and random dissolution. The lower the setting for credit availability increases the likelihood that firms with a higher competence level will be successful in seeking credit, whilst a low level in random dissolution implies that there is higher likelihood that firms with a lower financial stability level will become bankrupt. With these two variables in mind, the model was executed for five scenarios – both variables high, both low, one high and low and vice versa, and both set at medium levels.

Similar macro and micro analytical methods were used as in Stage 1. All GDP graphs were similar, indicating robust and growing economies. Cyclic behaviour was evident for all, an indicator that firms are entering and exiting the marketplace, although there was some instability in the frequency and amplitude of the cycles for the high credit availability and high dissolution scenario.

When evaluating the scenarios from a micro perspective, however, there are significant differences between the scenarios, and also a large difference in the number of effective firms formed in general as compared to Stage 1. Of the five scenarios, only two produced any effective firms – both of these scenarios had a high credit availability setting. This implies that conditions for effective firm formation were very difficult, with the combination of lower credit availability and the normal logic used for effective firm formation being an inhibition factor. The generally low number of effective firms formed even when credit availability was high is an indication of the difficulty imposed by this combination.

Table 10.2. Risk / Entrepreneur Competency Scenarios

Credit Availability Seed / Effective Firm Dissolution Seed Combinations	No. Effective Firms	Average Duration of Firm (Time Periods)	Average Net Account Level per Firm	Persistence of Profit (Average Net Account Level per Firm Duration)
100 (High) / 100 (High)	17	1,390	82,905	43
100 (High) / 100,000,000 (Low)	11	1,372	268,488	196

10.4.3 Stage 3 – Impact of Predator Behaviour

The mechanism used within the model for this behaviour is based on relative financial stability levels between firms. In general, the business rule allowing predator behaviour focuses on firms of both very high stability levels and very low. Once a firm has reached a high level they scan the environment for firms with levels below a set amount, then make a decision to acquire them. These decisions are affected by the inherent randomness settings within the model, being controlled by a Predator Behaviour Likelihood setting at the start of execution. The lower this Likelihood setting the more likely that predator behaviour occurs.

Macro outcomes with normal GDP over time were identical, economies apparently being robust and increasing generally with cyclic behaviour. The maximum GDP achieved in all cases was approximately 37,000.

Some micro-level analysis is outlined in Table 10.3. The similarity of results across multiple likelihood settings correlates with the outcomes from the macro GDP charts, emphasising a lack of sensitivity within the model for this type of behaviour.

However, some general trends are observable from the results. The higher the likelihood of predator behaviour the higher the number of effective firms being formed. Linked to these higher numbers, however, is a smaller duration. Average net account levels, or profit, and also persistence of profit is also less for the firms in the higher likelihood settings.

Table 10.3. Predator Behaviour Scenarios

Predator Behaviour Likelihood Settings	No. Effec- tive Firms	Average Duration of Firm (Time Periods)	Average Net Account Level per Firm	Persistence of Profit (Average Net Account Level per Firm Duration)
10 (Very High)	12	936	87,763	63
100 (High)	12	967	87,314	64
10,000 (Medium)	13	1,276	59,799	50
100,000 (Low)	13	1,276	59,799	50
100,000,000 (Very Low)	9	1,779	128,637	73

10.4.4 Stage 4 – Impact of R&D Investment

Innovation, with the consequent outcome of product improvements in the form of increased attractiveness to consumers, is the mechanism for firm competition in this stage. However, rather than innovation being a random event as in Stage 1, it is the outcome of deliberate investment in research and development. This type of innovation has been noted by Schumpeter as one form of economic growth (Kirchoff 1991).

Firms can undertake an investment in R&D, partaking some of their financial stability in the process, with an understanding that this investment may or may not be successful, i.e. their products may be improved as a result or they simply decrease their stability without any change in their product offering in the marketplace.

The model contains two environmental "levers" to explore the effect of planned innovation and the risk of consequent success or failure. These are run-time settings that represent the probability that a firm will make a strategic decision to invest in R&D, and the probability that these investments will be effective or not. The lower the values for these two settings, StategicDecisionMaking and Effectiveness, the greater the probability that they will occur for the firms selected as part of the random process inherent in the model. Investment can be classified as either proportional (to financial stability level) or fixed.

GDP profiles across all scenarios were similar, demonstrating cyclic growth as per past stages, however there were some anomalies. Very high and high combinations of investment and effectiveness (as seen in Figures 10.3 and 10.4), highlighting that excessive expenditure in R&D (even if succcessful) was counter-productive at an economy level.

Some micro-level analysis is summarised in tables 10.4 and 10.5. As per GDP outcomes, the10/10 scenario represents a poor outcome from both proportional and non-proportional perspectives. The medium/medium combination is a positive one for the number of effective firms and their duration but is the worst for net profit and profit persistence. The best scenarios for the com-

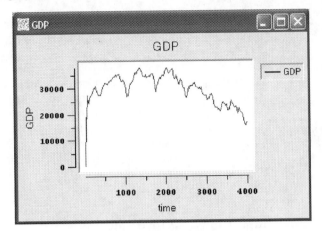

Fig. 10.3. HIGH Investment & Effectiveness (Non-Proportional)

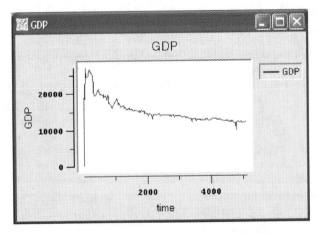

Fig. 10.4. VERY HIGH Investment & Effectiveness (Proportional)

Table 10.4. R&D Investment/Effectiveness Scenarios (Proportional)

R&D Investment & Effectiveness Combinations	No. Effective Firms	Average Duration of Firm (Time Periods)	Average Net Account Level per Firm	Persistence of Profit (Average Net Account Level per Firm Duration)
10 (V. High) / 10 (V.High)	6	667	244,827	166
100 (High) / 100 (High)	12	1,206	102,975	74
1000 (Med) / 1000 (Med)	12	1,658	89,241	13
10 (V. High) / 10,000 (Low)	12	1,143	153,207	84
10,000 (Low) / 10 (V.High)	12	1,546	134,579	41

Table 10.5. R&D Investment/Effectiveness Scenarios (Non-Proportional)

R&D Investment & Effectiveness Combinations	No. Effec- tive Firms	Average Duration of Firm (Time Periods)	Average Net Account Level per Firm	Persistence of Profit (Average Net Account Level per Firm Duration)
10 (V. High) / 10 (V.High)	9	407	-84,713	-234
100 (High) / 100 (High)	9	1,790	327,467	189
1000 (Med) / 1000 (Med)	12	1,658	89,241	13
10 (V. High) / 10,000 (Low)	8	593	94,228	89
10,000 (Low) / 10 (V.High)	12	1,546	134,579	41

bination of number of firms, duration, and general profit, are the high/high and very low/high combinations. Compared to the proportional scenario, the non-proportional high/low combination produces poor micro outcomes.

10.4.5 Stage 5 – Imitation and Marketing Defense Impacts

This last stage focuses product imitation and defense against imitation. It is similar in context to Stage 4 where firms make strategic choices as to where they should place their investments as a means of improving their competitive position (Schnaars 1994). As a form of innovation, imitation requires costs and capabilities to be present in a firm for the decision and action to occur (Andersen 2003).

Imitation is defined here as the action whereby a firm checks other firms for signs of products strength and increasing sales growth, then decides to invest in imitation such the firm's own product attractiveness is slightly higher than that of the other firm. Defense is defined as the action whereby a firm invests in its own product to increase its value and hence act as a defense against potential imitation. It should be noted that product attractiveness does not have to only imply that the product itself is superior to other products and hence becomes more attractive to consumers. The investment may be in the nature of a pure marketing investment such that the perception of the product is one of more attractiveness to the consumer – specific product features are only one element of the marketing mix.

From an environmental setting perspective, the variables ImitationInvest-ment and DefenseInvestment allow control over the degree of investment within the model. The lower the setting for these variables, the greater the probability that they will occur for the firms randomly selected. There were ten scenarios evaluated within this stage, five for proportional invest-ments and five for non-proportional. Within each group of five, the invest-ment/defense combinations were structured as high/high, medium/medium, low/low, high/low, and low/high.

GDP macro analysis revealed significant differences between each scenario, with proportional and non-proportional investment scenarios being similar across all combinations. All scenarios involving high or medium defense investment resulted in a declining economy. Only the two scenarios with low defense investment resulted in a robust economy, implying a clear link between declining economic strength and excessive defense investment.

Table 10.6. Imitation Defense Scenarios (Proportional Investment)

Imitation and Defense Investment Combinations	No. Effec- tive Firms	Average Duration of Firm (Time Periods)	Average Net Account Level per Firm	Persistence of Profit (Average Net Account Level per Firm Duration)
High(100)/High(100)	8	523	-104,611	-386
Med(10,000)/Med(10,000)	10	1,475	-37,049	-26
Low(1,000,000)/Low(1,000,000)	15	1,329	23,174	47
High(100)/Low(1,000,000)	9	1,521	-43,115	-109
Low(1,000,000)/High(100)	10	740	-80,403	-298

Analysis at a micro level tallies generally with GDP results, except in the area of a high/low combination. Whilst the economy appears robust in this scenario, negative profits are apparent for the firms involved. A situation where there is a small amount of competitive behaviour at both imitation and defense levels appears to result in the optimum conditions for both the economy and firms in general.

10.5 Conclusion

From the range of scenarios examined it was clear that varied run-time conditions produce quite diverse outcomes. This highlights the inherent sensitivity of the model, and the future potential for exploration within scenarios of multi-variable conditions.

Starting with the impact of random innovation, and incorporating the Product Attractiveness Level variable as a means of influencing consumer choice, the varied settings highlighted the dramatic impact that innovation (which is assumed to lead to changes in consumer product preferences) has on the economy and on firm attributes. The low innovation level setting indicated that there may be some tolerance level where a small amount is beneficial for the economy (from a maximum GDP perspective) and firm account levels.

In the area of credit availability where an Entrepreneur Competency Level variable was introduced as an aid in risk evaluation by banks, it was clear that availability has a significant impact on the number of effective firms formed. The scenario of high credit availability and low probability of dissolution produced the best result from a firm profit perspective but ran counter to expectations in that it contained a lesser number of effective firms.

Whilst the predator behaviour stage lacked sensitivity in its execution there was sufficient differentiation between a predator behaviour environment and a nominally non-predator behaviour environment. Generally the environment allowing this behaviour produced more effective firms, but these firms generally had less net profit. The financial stability level of firms (introduced as a variable to aid in firm decision-making in this stage and others) was somewhat better in a non-predator situation, but not by a significant amount.

In comparison with random innovation, the results for planned innovation as an expression of a competitive strategic decision are quite different. The use of R&D Investment and R&D effectiveness levers, as well as the decrease in financial stability as firm profit was reduced once a decision was made to invest in R&D, created a varied range of outcomes. Excessive levels of innovation for both types result in degradation of performance at an economy level, however the variance in the number of effective firms under different conditions of investment and effectiveness for strategic decision making circumstances is quite marked compared to the more linear outcomes from the random innovation stage. Medium investment/medium effectiveness combinations resulted in maximum firm numbers whilst several other combinations produce good profit outcomes and financial stability levels. No single scenario stood out, taking into account all measures available.

The last stage extended the decision making process within firms, involving the allocation of profits towards two types of competitive behaviour – imitation of another firm's product and a corresponding marketing-oriented defense against either real or potential imitation. The two variables involved in these decisions are Imitation Decision and Marketing Defense Decision. All decisions were focused on the improvement in a firm's Product Attractiveness Level, with a corresponding decrease in Financial Stability Level.

These imitation and defense scenarios demonstrated the widest variety in behaviour at GDP levels and effective firm formation, and produced the lowest profit levels than in any other stage. The low investment/low defense scenario was clearly the best across most measures.

References

[1] Andersen E.S.(2003) Schumpeterian games and innovation systems: Combining pioneers, adaptionists, imitators, complementers and mixers, Danish Research Unit for Industrial Dynamics (DRUID) Working Paper 2003-04-01
[2] Bloch, H.(2000) Schumpeter and Steindl on the dynamics of competition. Journal of Evolutionary Economics 10:343-353

[3] Bruun, C. & Luna, F.(2000) Endogenous Growth with Cycles in a Swarm Economy: Fighting Time, Space, and Complexity, Economic Simulations in Swarm: Agent-Based Modelling and Object Oriented Programming, Luna, F. & Stefannson, B. (eds.), Kluwer

[4] Carlsson, B. & Eliasson, G.(2001) Industrial Dynamics and Endogenous Growth. Danish Research Unit for Industrial Dynamics (DRUID) Working Paper 2001-05-29

[5] Clemence, R.V. & Doody, F.S.(1966) The Schumpeterian System, Augustus Kelley, New York

[6] Jacobson, R.& Hansen, G.(2001) Modeling the Competitive Process. Managerial and Decision Economics 22: 251-263

[7] Kirchoff, B.A.(1991) Entrepreneurship's Contribution to Economics. In: Winter (ed.) Entrepreneurship: Theory and Practice.

[8] Kirzner, I.M.(1973) Competition and Entrepreneurship. University of Chicago Press

[9] Knight, F.H.(1921) Risk, Uncertainty, and Profit. Library of Economics and Liberty, III.IX.7

[10] Metcalfe, J.S.(1998) Evolutionary Economics and Creative Destruction. Routledge , New York

[11] Nelson, R.R.(1994) Economic Growth via the Coevolution of Technology and Institutions. In: Leydesdorff, L. & van den Besselaar, P. (eds.) Evolutionary Economics and Chaos Theory. Pinter (London)

[12] Schnaars, S.P.(1994) Managing Imitation Strategies. The Free Press, New York

[13] Shionoya, Y.(1997) Schumpeter and the idea of social science, Cambridge

Part IV

Social Interaction - Network Effects

The Wisdom of Networked Evolving Agents

Akira Namatame

Department of Computer Science, National Defense Academy, Yokosuka, Kanagawa, 239-8686, Japan, nama@nda.ac.jp

Summary. The fact that selfish behavior may not achieve full efficiency at the aggregate level has been well known in the literature. Therefore we need to cope with the socio-economic system by attempting to stack the deck in such a way that individuals with selfish incentives have to do what is the desirable thing. Of particular interests is the question how social interactions among individuals can be restructured so that they are free to choose their actions while avoiding outcomes that none would have chosen. In this paper, we study the collective construction process of social norms and the emergence of collective intelligence of networked evolving agents. The wisdom of collective agents is interpreted as emergence of behavioral rules that constitute constraints on social interactions so that self-interested agents can achieve efficient and equitable outcomes.

11.1 Introduction

It is common in many markets that the buying decision of one consumer influences the decisions of others. The general effects applying to all the consumer decisions, markets also have strong positive or negative network effects. Popular examples of positive network effects are the willingness to adopt a product innovation correlates positively with the number of existing adopters. Positive network effects in markets mainly originate from two different areas, the need for compatibility to exchange information and the need for complementary products and services.

On the other hand, in many cases, the existence of network externalities results in so called madness of crowd. Economic implications resulting from the bandwagon and herding behavior are broadly discussed in the literature [1]. Network externalities often lead to Pareto-inferior outcomes due to coordination failure. The fact that selfish behavior may not achieve full efficiency at the aggregate level has been also known in the literature. Recent research efforts have focused on quantifying the loss of system performance due to selfish and uncoordinated behavior. The degree of efficiency loss is defined as the price of anarchy [17]. The reason why uncoordinated activities of agents pursuing their own interests often produce outcomes that all would seek to avoid is that each agent's behavior affect the others and these effects are often not included in whatever optimizing process made by other agents. These unaccounted effects on others are called network externalities

Socio-economic systems consist of individuals and the socio-economic system in which they interact. On the other side, a collective of individuals creates the socio-economic system of which they are parts. Therefore, the essence of the socio-economic system is that it is the individuals who are making their own decisions. We need to cope with the socio-economic system by attempting to stack the deck in such a way that individuals have selfish incentives to do what is the desirable thing. Explicit or implicit coordination is necessary to achieve individuals' goals more efficiently. However, many aggregate social outcomes have emergent properties that cannot be trivially derived from the properties of the members who consist the socio-economic system.

In his book, titled The Wisdom Of Crowds, Surowiecki explores an idea that has profound implications: a large collection of people are smarter than an elite few, no matter how they are brilliant and better at solving problems, fostering innovation, coming to wise decisions, even predicting the future [18]. He explains the wisdom of crowds emerges only under the right conditions: (1) diversity, (2) independence, (3) decentralization, and (4) aggregation. His counterintuitive notion, rather than the madness of crowd such as herding, cascade as traditionally understood [1], suggests new insights for the issue on how complex social and economic activities should be organized.

This observation derives requirements for a more general model of network effects. Therefore a new area of research is emerged aiming at explaining the phenomena of strong positive or negative network effects in markets and their implications on market coordination and efficiency. However, the assumptions and simplifications implicitly used for modeling social interaction processes fail to explain the individual cognitive decision-making process as well as the network structure. A crucial ingredient in social interaction models is the network structure in which individuals interact.

Many spheres of social interactions are governed by social norms such as reciprocity and equity. Social norms are self-enforcing patterns of social behavior. It is in everyone's interests to conform given the expectation that others are going to conform. It is a rule of the action choice that assigns a rule to each agent that is an optimal in the sense no one has an incentive to deviate from it. Although social norms can potentially serve useful constructs to understand human behavior, there is little theory on collective construction of social norm.

Epstein and Axtell work on the evolutionary process that brings about norms [8]. They work on the model and discover that, once people got the norm they are no longer trying to make decisions the way they make them before there was a norm. Once people have norm they can internalize the norm, they can remember the norm, and they can teach the norm. If we want to see whether game theory can be of any help in thinking about which norms come about, how they come about, how durable they may be when they come about, then game theory can help.

We study the collective construction process of social norms as the wisdom of networked agents. The social space consists of networks of self-interested agents, continuous evaluations of their performance as well as their behavioral rules. Behavioral rules are here treated as the constraints on individual action and they specify the action choice based on the specific outcomes. The learning of new behavioral rule, and the strife of each agent to act in keeping with the coupling with the neighbors constitute the collective construction of social norms. Social norms are here treated as the shared behavioral rules that constitute common constraints on all individuals in a society. For agents in a social context to achieve collective intelligence, it is a continuous process that requires social behavior based on social rationality [14]. To in turn achieve social rationality requires for individually rational behavior to be constrained by some obligations. We also study collective construction of social norms by focusing on the relation between micro and macro levels of constraints on the evolution of socially intelligent behavior.

11.2 Game Theoretic Models of Social Interaction

The interaction structure specifies who affects whom, and this network structure may vary from one individual to another. Social interdependence can be understood as a dependence of outcomes of one individual on another individual behavior. Such a relationship between payoffs for choices of different individuals is usually described with the formalism of the game theory.

11.2.1 Coordination Game: Nash Demand Game

We begin by modeling a bargaining process between two agents. Consider two agents, A and B, each of whom demands some portion of a "pie", which we take as a metaphor for a piece of available resources which is divisible. A way of modeling this bargaining situation is the Nash demand game: each agent gets his demand if the sum of the two demands is not more than 10; otherwise each gets nothing. We simply assume that each agent can make just three possible demands: low (3), medium (5), and high (7). The payoffs (in share) from all combinations of demands are shown in Table 11.1. This Nash demand game yields a coordination game in which there are three pure Nash equilibria: (S_1, S_3), (S_2, S_2), and (S_3, S_1).

Axtell and his colleagues explored which equilibrium emerges at the aggregate level from the repeated pair-wise interactions of self-interested agents [2]. They consider a population of N agents and in each match, one pair of agents is drawn at random from the, and they play the Nash demand game in Table 11.1. Therefore there is no network effect. Each agent makes a demand that maximizes her expected payoff (best-response) about the opponent's behavior, which defines the current state of the population. However, with some small probability, each agent chooses one of the three demands at random. They showed that the equity norm, corresponding to the equilibrium (S_2, S_2), and each agent demand medium has a large basin of attraction. In the terminology of evolutionary game theory, the equity norm is stochastically stable. Occasional random choices create noise in the evolutionary dynamics, which implies that no state is perfectly absorbing. However, there two regions of the state space-one equitable, the other fractious-that every persistent: once the process enters such a region, it tends to stay there for a long period time. Therefore they showed that the emergence of the equity norm by self-interested agents is hard in the sense that it takes exponential time to achieve it from some initial states.

Table 11.1. The payoff matrix of the Nash demand game

AgentB AgentA	S_1 (High)	S_2 (Medium)	S_3 (Low)
S_1 (High)	0 0	0 0	3 7
S_2 (Medium)	0 0	0 5	3 5
S_3 (Low)	7 3	5 3	3 3

11.2.2 Dispersion Game: Generalized-Rock-Scissors Paper Game

The hand game "Rock-Scissors-Paper (RSP)" is also known as "Janken" in Japan, has been around the world for a long time. It most often used to solve small conflicting matters between peoples but it can also be played to decide larger matters, as part of tournament, our simply as a diversion. The basics of the game consist of each player shaking a fist a number of times and then extending the same hand in a fist ("rock"), out flat ("paper"), or with the index and middle fingers extended ("scissors").

The RSP game is also important for the study in many ecosystems. Kerr *et al* set out to investigate the mechanisms that maintain biodiversity in ecosystems [10]. Studies of three bacterial strains engaged in an interaction that mimics the game, rock-scissors-paper, show the importance of localized interactions in maintaining biodiversity. Kerr and colleagues are not the first to show that localized interactions of the rock-scissors-paper type can turn a one winner- outcome into a dynamic coexistence of all three types, endlessly chasing each other across the board [21].

We consider a population of agents located a lattice network repeatedly play the generalized RSP games with the payoff matrix in Table 11.2. The generalized RSP game in Table 11.2 has the unique Nash equilibrium, and each strategy, rock, scissors and paper should be selected with the same probability $1/3$. The expected payoff of each agent with this mixed Nash equilibrium is $(\lambda + 1)/3$. If the parameter λ is greater than 2, the payoff at Nash equilibrium is asymmetric, and then the problem of the efficiency and fairness and may arise.

Table 11.2. The payoff matrix of the generalized rock-scissors-paper game $(\lambda \geq 2)$

AgentA \ AgentB	S_1 (Rock)		S_2 (Scissors)		S_3 (Paper)	
S_1 (Rock)		1		0		λ
	1		λ		0	
S_2 (Scissors)		λ		1		0
	0		1		λ	
S_3 (Paper)		0		λ		1
	λ		0		1	

11.3 Strategy Choice Based on Learnable Behavioral Rule

In orthodox rational choice theory, agents are modeled as cognitively sophisticated and entirely self-interested decision makers who evaluate every future consequence of possible actions and select the action alternative that maximizes own payoffs. Discrete choice analysis grounded in the theory of utility maximization has proven quite successful in terms of its usefulness. However, this approach is being challenged by a line of research originating in cognitive psychology that is causing economists to re-examine the standard model of choice behavior [13]. In the words of the psychologist Kahneman, economists have preferences; psychologists have attitudes [20].

However, experimental evidence supports the view the behavioral rules are the proximate drivers of most human behavior. The rule-governed action can be also pictured as a quasi-legal process of constructing a satisfying interpretation of the choice situation. The behavioral rules we do use are essentially defensive ones, protecting us from mistakes that perceptual illusions may induce. However, the question remains as to whether behavioral rules themselves develop in patterns that are broadly consistent with the rational model postulates. This is a vital scientific concern. If there are preferences behind the formation of behavioral rules, then how they are correlated with these underlying preferences.

We seldom do new things. Most behaviors are repeated, but many researchers do not pay much attention to this aspect. Few would dispute the claim that most behaviors are repetitive, yet in spite of a large literature on learning, the habit concept has received only minor attention. The sort of coordination problems we have in mind are those that we commonly solve without thought or discussion-usually so smoothly and effortlessly that we don't even notice that there is a coordination problem to be solved.

Verplanken and Aarts define habits as learned sequences of acts that have become automatic responses to specific cues, and are functional in obtaining certain goals or end-states [20]. Obviously, many behaviors may fall under this definition, varying from being very simple to being complex. Habits are learned sequences of actions. Habits are also automatic responses to specific cues. Habitual acts are also instigated as immediate responses to specific situations. These responses occur without purposeful thinking or reflection and often without any sense of awareness. Most habits are created and maintained under the influence of learning. For instance, behavioral rule that has positive consequences is more likely to be repeated, whereas negative consequences make repetition less likely. Repeated behaviors may turn into habits, which are automatic responses to specific cues and are functional in obtaining certain goals. We may want new and desired behaviors to become habits, which makes them stable and difficult to change. Habituation thus become a behavioral rule.

Social norms and habit influence in turn individuals' purposive behaviors based on their current preferences. This bi-directional causal relationship is at the essence of the study of the cognitive decision-making process. Understanding the nature of the relationship between two different levels at which actual choice is a grand challenge of cognitive science. Explanation of this relationship calls for examining the types of social interactions that link individuals in social contexts.

In this paper we propose a hybrid choice model based on both rule-based choice and preference-based choice as shown in Figure 11.1. Agents adhere to behavior rules via local adaptation of behavior. The adaptation of behavior rules consists of an internalization of social norms, or more precisely a synchronization of the individual behavioral rule to those of the other neighbors. Each agent applies the hybrid choice model based on both agent specific assessments of the situations (rational choice model) and social norms or habits (rule based choice model). Social norms have been treated here as constraints on agent-specific rational choices. Each agent is modeled to evolve her behavioral rule. This hybrid choice model at individual levels is the core of emergent socially intelligent behavior.

We stresses that the performance of the socio-economic system consisting of self-fish agents depends on how they are properly coupled. A strategy choice based on the behavioral rule for repeated play of the game uses the recent history of play to choose one of the three strategies for the next play. Here, we assume that each agent can refer to the last outcome. Each behavioral rule is represented as a binary string so that the genetic operators can be applied. We represent a behavioral rule by a 3-bits string using $S_1=0$, $S_2=1$ and $S_3=2$.

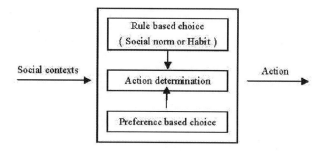

Fig. 11.1. A hybrid choice model

In order to accomplish this we use a bit string. Since no memory exists at the start of the game, extra one bit is needed to specify a hypothetical history at the beginning.

Each position p_j, $j=1,..,12$, in Figure 11.2 represents as follows. The first position p_1 encodes the initial strategy that the agent takes at each generation. A position p_j, $j = 2, 3$, encodes the history of mutual hands (rock, scissors, or paper) that agent and her opponent took at the previous round. A position p_j, $j = 4, ., 14$, encodes the action that the agent takes corresponding to the values at the positions p_j, $j = 2, 3$.

There are nine possible outcomes for each round. We can fully describe a behavioral rule by recording what the strategy will do in each of the nine different outcomes that arise in the last play of the game. A rule must specify depending each outcome, what strategy the agent should choose at the net round. Since there are three strategies, the number of possible behavioral rules is 3^9. The hope is that agents would find a better behavioral rule out of the overwhelming possible rules after a reasonable number of plays.

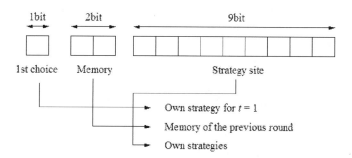

Fig. 11.2. An agent's memory of a behavioral rule

11.4 Individual Learning vs. Social Learning

To achieve desirable outcomes, a primary question is how each individual should learn in the context of many learners [22]. There are two competing approaches for describing the learning model of the population: the microscopic model based on individual learning and the macroscopic model based on social learning.

In the category of individual learning, agents are modeled to have some repertories of behavioral rules, and they update those rules using the existing rules within as shown in Figure 11.3(a). Natural selection operates on the local probability distribution of behavioral rules within the repertoire of each individual agent. In an individual learning model, we could say that each agent checks if another randomly chosen agent in the population gets a higher payoff, and, if so, switches to that behavior with a probability proportional to the payoff difference.

There is no imitation or exchange their experiences among agents in individual learning. On the other hand, social learning becomes valuable in a social context, since it can help to surface new ideas and generate social consensus on issues that no single individual can effectively make right decision about alone. Social learning can also be extended beyond the boundaries of a single agent. Social learning is one that has an internal process for cultivating individual learning and connecting it to others. So when faced with change, a collective has the requisite energy and flexibility to move in the direction it desires.

In an orthodox social learning model as shown in Figure 11.3(b), agents play based on the prescribed behavioral rules. The summed payoff of each game provides the agent's fitness. After every individual has played the game with her neighbors, each rule of the agents is updated according to the general evolutionary rules, and the behavioral rule is crossover with the most successful behavioral rule of her neighbors. Their success depends in large part on how well they learn from their neighbors. If an agent gains more payoff than her neighbor, there is a chance her behavioral rule will be imitated by others.

The principle of social learning itself can be thought of as the consequence of any one of three different mechanisms. It could be that the more effective individuals are more likely to survive and reproduce. A second interpretation is that agents learn by trial and error, keeping effective rules and altering ones that turn out poorly. A third interpretation is that agents observe each other, and those with poor performance tend to imitate the rules of those they see doing better.

The most unrealistic aspect of the rule learning is the large number of strategies each agent considers. Even if the set of rules is limited to very simple ones, each agent remembers to many strategies. A realistic model should account for the fact that agents consider a much smaller number of rules from which they learn and make decisions; and that the rules agents consider are

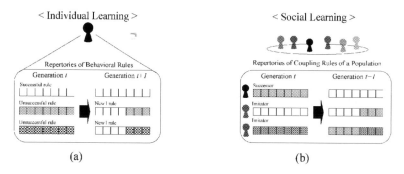

Fig. 11.3. Individual learning vs. social learning for evolving behavioral rules.

often preconditioned by factors such as imitation that have evolved over the generations.

11.5 Social Interactions of Networked Agents

A crucial ingredient in social interaction models is the network structure in which individuals interact. The interaction structure specifies who affects whom, and this network structure may vary from one individual to another. The agents involved would learn two things: with whom to interact and how to behave. That is to say that learning dynamics operates both on network structure and strategy. The interaction structure specifies who affects whom.

In order to describe the ways of interaction, the random matching model is frequently used. In the random matching model, in which each agent is assumed to interact with a randomly chosen agent from the population. There are also a variety of interaction models, depending on how agents meet, and what information is revealed before interaction.

There are many situations in which a spatial environment becomes a more realistic representation, since interactions in real life rarely happen on such a macro-scale as assumed in the global interaction model. Spatial interaction is generally modeled through the use of the two dimensional (2D) grid in Figure 11.4(a) with each agent inhabiting each cell of the lattice on the grid. Interaction between agents is restricted to nearest neighboring agents. Each agent chooses an optimal strategy based on local information about what her neighbors will choose. However, the consequences of their choices may take some time to have an effect on agents with whom they are not directly linked.

At another end of the spectrum we have models where individuals interact with both fixed their neighbors and randomly chosen agents from the population. Watts and Storogatz introduced a small-world network architecture that transforms from a coupled system with nearest neighbors to a randomly coupled network by rewiring the links between the nodes [21]. For instance, consider a two-lattice model in which each node is coupled with its nearest neighbors, as shown in Figure 11.4(b)(c). If one rewires the links between the nodes with a small probability, then the local structure of the network remains nearly intact.

If we fix the interaction structure, we get models of the evolution of strategies in games played on a fixed network structure. An interaction structure need not be deterministic. In general, it can be thought of as a specification of the probabilities of interaction with other individuals. By far the most frequently studied interaction structure is one in which the group of individuals is large and individuals interact at random. That is to say that each individual has equal probability of interacting with every other individual in the population.

Some researchers also concern the impact of different network structures on equilibrium selection in the context of coordination games. If agents can choose their partners to interact, then they will form networks that lead to play of the efficient Nash equilibrium in the underlying coordination game. Ellison analyzed the role of local interactions for the spread of particular strategies in coordination games, showing, how play converges to the risk-dominant equilibrium if agents are located on a circle and interact with their two nearest neighbors [7]. Similarly, Blume and Kosfel proved the convergence to the risk-dominant equilibrium in a population of agents located on a two-dimensional lattice [4][11]. Kuperman and Abramson studied an evolutionary version of the prisoner's dilemma game, played by agents placed in a small-world network [12]. Agents are able to change their strategy, imitating that of the most successful neighbor. They found that collective behaviors corresponding to the small-world network enhances defection where cooperation is the norm in the fixed regular network.

Ultimate interest resides in the general case where structure and strategy co-evolve. These may be modified by the same or different kinds of learning. They may proceed at the same rate or different rates. The case where structure dynamics is slow and strategy dynamics is fast may approximate more familiar models where strategies evolve on a fixed interaction structure. Whether co-evolution of structure and strategy supports or reverses the conventional wisdom about equilibrium selection in this game, depends on the nature and relative rates of the two learning processes.

11.6 Simulation Results

11.6.1 Nash Demand Game

Figure 11.5 shows (i) the average payoff per agent and (ii) the ratio of each strategy over generation when each agent repeatedly plays the Nash demand game with the payoff matrix in Table 11.1. After the 12 generation, the average payoff is increased to 5, and every agent chooses S_2 (medium). Many spheres

Fig. 11.4. Social interactions of networked agents. (a) p = 0, a regular lattice. (b) p= 0.1, some of the links have been re-wired resulting in a small–world network. (c) p = 0.5, additional re-wiring has occurred. As p approaches 1, a transition to a random network will occur.

of social interactions are governed by social norms such equity. In this case, the efficient and equity norm is emerged over the networked agents.

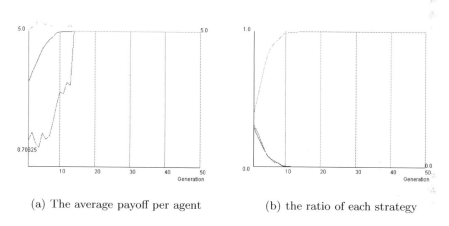

(a) The average payoff per agent   (b) the ratio of each strategy

Fig. 11.5. Simulation results of Nash demand games

11.6.2 Generalized RSP Games

We simulated several cases by changing the parameter value ofλ. We also consider the effect of implementation error. That is, there is small probability of choosing the different strategy from the one specified by the rule. Significant differences will be observed when agents have small chances of making mistakes.

(Case 1) $\lambda = 2$: Figure 11.6 shows (i) the average payoff per agent and (ii) the ratio of each strategy over generation when we set $\lambda = 2$in Table 11.2. All agents receive the same average payoff if $\lambda=2$ by choosing S_2 (scissors).

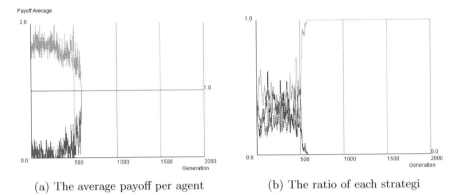

(a) The average payoff per agent   (b) The ratio of each strategi

Fig. 11.6. Simulation results with $\lambda = 2$

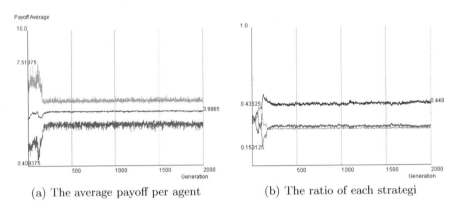

(a) The average payoff per agent   (b) The ratio of each strategi

Fig. 11.7. Simulation results with $\lambda = 10$ implementation error: 10%

(Case 2) $\lambda = 10$. We now investigate the strategic situation by increasing the payoff of winning the game by setting λ=10 in Table 11.3. Figure 11.7 shows the simulation results with the implementation error of 10%. Figure 11.7(a) shows the payoff per agent at each generation. The average payoff per agent is approximately 3.9, which is higher than the expected payoff at Nash equilibrium, which is approximately 3.7. Figure 11.7(b) shows the strategy distribution, and which eventually converges to the same distribution with Nash equilibrium. In the beginning, 400 different coupling rules were aggregated into three types, as shown in Table 11.3.

The game between two agents who play with the behavioral rules can be described as a stationary stochastic process. The state transition of the outcomes when both agents choose their strategies according to the same coupling rule of type 1 is illustrated in Figure 11.8 as the state transition diagram.

(Case 1) Agents who have the same behavioral rule:

The strategy choices between two agents with the same coupling rule type i, $i = 1,2,..,8$, are shown in Figure 11.8(a). In this figure, there is one absorbing state at "22" and one limiting cycle. The state diagram contains two paths, one for moving towards to the absorbing state and one for the limiting cycle, and there is no path between the two cycles. As shown in Table 11.3, agents also learn to initiate the play by choosing "paper (2)" and strategy choices eventually converge to "22". This means that if an agent plays with other agents of the same rule, they converge to the state of a tie, and receive the lower payoff of 1.

(Case 2) Agents who have the different behavioral rules:

We now investigate the state diagrams of plays by two agents who have different coupling rules in Table 11.3. The state diagram is shown in Figure 11.8(b). If the system were to start from the set of the states, it would evolve to an attractor. These are known as the basin of attraction. In this case, the point attractor for the state of the systems is replaced by a circle, and in the limit, the system moves endlessly around this circle. Starting from any state, it eventually converges to an efficient cycle such that agents win three times and lose three times.

In the framework of the hybrid choice model as shown in Figure 11.1, agents, facing different social contexts, are assumed to choose their actions randomly with some small probability. In this simulation, we simply assume that agents have a chance to choose the other strategy from the strategy specified by rule. We model this process by introducing some mistakes or implementation errors.

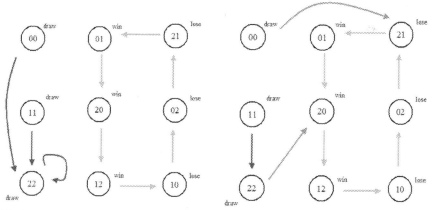

(a) Agents who have the same behavioral rules

(b) Agents who have the different behavioral rules

Fig. 11.8. The state diagram of play between two agents

With some mistakes, some interesting properties emerge. If an agent A chooses rock (0) and her opponent Bchooses "scissors (1)" (in this case she wins and her opponent loses), then in the next round agent A chooses "scissors (1)" and agent Bchooses "paper (2)". In the following round agent Achooses "paper (2)" and agent B chooses "rock (0)". Therefore, agent Awins three times and agent B loses three times. However, after these games, the two agents completely reverse roles, and the winning agent thus far, agent A, chooses "scissors (1)" and the losing agent thus far, agent B, chooses "rock (0)". After these three one-sided games, they trade places. The winner thus far chooses "scissors (1)" and loser thus far chooses "rock (0)". The winner then becomes the loser, and vice versa. In total, her opponent wins three times. Both agents are eventually absorbed into the limit cycle of the three-wins and three-losses. Thus far, this agent wins three times and her opponent loses three times. Therefore, the two agents switch roles as winner and loser. Since both agents win three times and lose three times, on the average, they gain the payoff at Pareto-efficiency.

Table 11.3. Learnt behavioral rules by 400 agents

Rule type	Initial Strategy	Strategy site									Number of agents
	1	2	3	4	5	6	7	8	9	10	
1	2	0	1	0	2	0	2	1	0	0	149
2	2	2	1	0	2	0	2	1	0	0	102
3	2	0	1	0	2	2	2	1	0	0	58
4	2	2	1	0	2	2	2	1	0	0	41
5	2	2	1	0	2	0	2	1	0	2	20
6	2	0	1	0	2	0	2	1	0	2	15
7	2	0	1	0	2	2	2	1	0	2	9
8	2	2	1	0	2	2	2	1	0	2	6

11.7 Social Norms Emerged over Networked Agents

Many laboratory experiments and field observations indicate that humans are social animals who take a strong interest in the effects of their actions on others and whose behavior is not always explained by simple models of selfish behavior. Reciprocity and the presence of other social norms can support a great deal of social intelligent behavior [14]. The simulation results in the previous section implicate that we have a tools for examination how social norms evolve in a society that begins in an amorphous state where there is no established common behavioral rules and individuals only rely on hearsay to determine what to do.

We discussed how desirable social norms emerge in a society of interacting agents. Especially, we investigate the conditions under which the norm will emerge and dominate in various social setting, and discuss the importance of collective norm construction for evolving and sustaining a desirable society of efficiency and equity. After collective construction of social norms, there is no need to assume a rational calculation to identify the effective behavioral rule. Instead, the analysis of what is chosen at any specific time is based upon an implementation of the idea that effective behavioral rules are more likely to be retained than ineffective ones.

Epstein and Axtell extended the literature on the evolution of norms with an agent-based model [8]. In their model, agents learn how to behave (what norm to adopt), but they also learn how much to think about how to behave. The point of their model is that many social norms or conventions have two features of interest. First, they are self-enforcing behavioral regularities. But second, once entrenched, we conform without thinking about it. Indeed, this is one reason why social norms are useful; they obviate the need for a lot of individual computing.

The asymmetry in payoffs from interaction induces agents to learn the behavioral rule that breaks the asymmetry. Hanaki used adaptive models to understand the dynamics that lead to efficient and fair outcomes in a repeated battle of the sexes game [9]. He develops a model that not only uses reinforcement learning but also the evolutionary learning that operates through evolutionary selection. He found that the efficient and fair outcome emerges relatively quickly through turn taking. However, his model requires a long run pre-experimental phase before it is ready to take turn. Turn taking in the battle of the sexes game is just one of many game theoretic phenomena, and it raises an important general point for further studies.

Browning and Colman also investigated how coordinated, alternating cooperation can evolve without any communication between agents who play battle of the sexes game [5]. They study the nature, properties and phenomena of coordinated alternating cooperation in a range of dispersion games with asymmetric equilibria. By alternating coordination the agents benefit from it, however, how agents evolves alternating coordination without communication is not fully explained.

We consider the generalized RSP game in which favourable payoffs are possible only if one agent acts one way while the other acts the opposite way. To coordinate successfully, the agents have to alternate or take turns, out of phase with each other. If this type of social interaction is repeated, the agents benefit by coordinated alternation by taking turns in choosing one of the three strategies and there is evidence to show that this type of turn-taking occurs quite commonly in nature. Give-and-take is a strategy that is intuitive and simple, but even so it is beyond the scope of most traditional learning models [15]

11.8 Conclusion

Networks of evolving agent are likely to foster social interactions where individual self-interest is consistent with behavior that maximizes the social welfare. Social interaction in such network structure is best modeled as a repeated game. In repeated games, where an agent's actions can be observed and remembered by other agents, almost any pattern of individual behavior, including behavior that maximizes the collective payoff, can be sustained by social norms that include obligations to punish norm violations by others. Where many equilibria are possible, collective construction of social norms is likely to play a major role in determining Pareto-efficient equilibrium will obtain.

We analyzed the emergence of socially intelligent behavior when evolving agents are networked in social spaces. Our problem is to explain how such socially intelligent behaviour could have evolved, given that natural selection operates at the individual level. The framework of collective evolution distinguishes from the concept of co-evolution in three aspects. First, there is the coupling rule: a deterministic process that links past outcomes with future behavior. The second aspect, which is distinguished from individual learning, is that agents may wish to optimize the outcome of the joint actions. The third aspect is to describe how a coupling rule should be improved with the criterion of performance to evaluate how the rule is doing well.

The performance assessment at the individual levels gradually evolves, in order for the agent to act in accordance with the behaviors of her neighbors. Social norms are not merely the union of the local behavioral rules of all agents, but rather evolve interactively, as do the local behavioral rules of the agents. In an evolutionary approach, there is no need to assume a rational calculation to identify the best behavioral rule. Instead, the analysis of what is chosen at any specific time is based upon an implementation of the idea that effective behavioral rules are more likely to be retained than ineffective ones.

References

[1] Adamatzky, A. (2005). Dynamics of Crowd-Minds, World Scientific
[2] Axtell,R and Epstein,M, and Young, P. (2001). The emergence of classes in a multi-aget bargaining model in. Social Dynamics, , Durlauf, N. and Young, P (eds). Brookings Institution Press, pp.191-211
[3] Bergstrom, T. (2002) Evolution of Social Behavior: Individual and Group Selection, Journal of Economic Perspectives, Volume 16, pp. 67-88
[4] Blume, E. (1993). The statistical mechanics of strategic interaction, Games and Economic Behavior, 5, pp. 387-424
[5] Browning, L.and Colman, M. (2004). Evolution of coordinated alternating reciprocity in repeated dyadic games. Journal of Theoretical Biology 229,pp 549-557

[6] Dick, S, Krambeck, M, and Milinski, S (2003) Volunteering leads to rock-paper-scissors dynamics in a public goods game, Nature. Vol. 425, pp390-393

[7] Ellison, G. (1993). Learning local interaction, and coordination, Econometrica,61, pp.1047-1071

[8] Epstein, J. M. and Axtell, R. (1996). Learning to be thoughless: Social norms and individual computation, Working paper no.6, Brookings Institution, 1999

[9] Hanaki, NCSethi, R, ErevDI and Peterhans. L (2005). Learning Strategies. Journal of Economic Behavior and Organization, Vol. 56, pp.523-542

[10] Kerr, B, Riley, M, Feldman, Brendan, M. and Bohannan, M. (2002). Local dispersal promotes biodiversity in a real-life game of rock-paper-scissors, Nature, Vol. 418, pp 171-174

[11] Kosfeld, M. (2002). Stochastic strategy adjustment in coordination games, Economic Theory, 20, pp. 321-339

[12] Kuperman and Abramson (2001), Social games in a social network, Phys. Rev. E 63

[13] Manski, C. F. and McFadden, D. L. (Eds) (1981). Structural Analysis of Discrete Data and Econometric Applications, The MIT Press

[14] McMahon,C. (2003). Collective Rationality and Collective Reasoning, Cambridge University Press

[15] Namatame, A. (2006) Adaptation and Evolution in Collective Systems, World Scientific

[16] Nowak, M. A. and Sigmund, K. (2004). Evolutionary dynamics of biological games, Science, 303, pp. 793-799

[17] Roughgarden, T. (2005). Selfish Routing and the Price of Anarchy, The MIT Press

[18] Surowiecki, J. (2004). The Wisdom of Crowds, Random House

[19] Skyrms, B and Pemantleá, R. (2000) A dynamic model of social network formation, PNAS, vol. 97, pp.9340-9346

[20] Verplanken B, Aarts H (1999). Habit, attitude, and planned behavior: is habit an empty construct or an interesting case of goal-directed automaticity? European Review of Social Psychology, Vol.10, pp.101-134

[21] Watts, D. and Strogatz, H. (1998). Collective dynamics of small-world networks, Nature, 393, pp. 440-442

[22] Young, H. P. (2005). Strategic Learning and Its Limits, Oxford Univ. Press.

12

Artificial Multi-Agent Stock Markets: Simple Strategies, Complex Outcomes

A.O.I. Hoffmann, S.A. Delre, J.H. von Eije, and W. Jager

University of Groningen, Faculty of Economics, Landleven 5, 9700 AV Groningen, The Netherlands, a.o.i.hoffmann@rug.nl

12.1 Introduction

Both micro level investor behavior as well as macro level stock market dynamics are research fields that are full of "puzzles" or unresolved research questions and therefore enjoy a strong interest of scholars and practitioners alike. On a micro level, aberrances in individual investor behavior are the subject of intense debate in e.g., behavioral finance (for an introductory overview of the field, see e.g. Nofsinger 2002; Schleifer 2000; Shefrin 2002; Shiller 2005). On a macro level, the absence of (linear) autocorrelation, and the occurrence of fat tails and volatility clustering in asset returns distributions are often studied stylized facts (Cont 2001).

Methodologically, there is a great heterogeneity in the techniques used to solve the above-mentioned puzzles. Surveys, case studies, laboratory experiments, and a plethora of statistical analysis are amongst the many methods that are used in this field. A relatively recent development in finance is the use of multi-agent simulation models as a research method (LeBaron 2000, 2005). The usage of multi-agent simulation models allows researchers both to make a coupling between the before identified micro and macro levels and to get a better understanding of the complexity that is often experienced in this field.

Investor behavior and related stock market dynamics are fields par excellence to observe complexity. Often, macro level outcomes, such as crashes and bubbles, "emerge". Interaction and nonlinearity in the micro level behavior of actors may cause these emergent phenomena. Small changes in the initial situation of a model or in the behavior of one or several interacting actors may lead to completely different outcomes on a macro level. Social simulation is a particularly appropriate tool in helping to explain the interactions between the micro and macro level of this complex behavior.

12.2 Background

A first step when using multi-agent simulation research to solve puzzles in finance is to formalize a limited number of micro level agent rules that, in the ideal situation, represent empirically found characteristics of investors' behavior. A population of investor agents is generated by the simulation model and these agents are provided with these rules. Subsequently, a number of simulation experiments are performed and finally, the macro level results (often in the form of stock price or returns time series) of these simulation experiments are compared with data from real stock markets in order to see to what extent real-life stylized facts are replicated.

In this paper, we continue the line of research of Hoffmann, Delre, Von Eije, & Jager (2005). In that paper, the need to incorporate theories of social needs, social interactions and social networks of investors in finance research - as first introduced in Hoffmann and Jager (2005) - was argued for. The objective of the research program is to identify critical micro level factors that drive investors' behavior and to explain complex macro level phenomena that result from the aggregation and interaction of micro level investor behavior. An adapted version of the model of Day & Huang (1990) is explored, which can be seen as a simple nonlinear dynamical system. The power of this simple model resides in the fact that simple agent rules are able to generate non-linear dynamics like stock market price and returns time series. Without any news, e.g., in the form of noise, this model is able to capture a number of stylized facts that are often observed in financial markets, like volatility clustering. The interactions between fundamental and trend following agents alone is enough to generate these complex outcomes. In the next section, the model will be briefly described.

12.3 The Simulation Model

In the model, investors can follow either a more fundamentally based "rational" strategy (called the α-strategy) or a more socially based trend following strategy (called the β-strategy). The α-strategy is based on a comparison between the current market price p and a given long-run investment value u. Whenever the market price is below the long-run investment value, the α-investor buys. Whenever the market price is above the long-run investment value, the α-investor sells. When the market price equals the long-run investment value, the α-investor holds. This behavior is limited by a topping price M (set at 1.0) and a bottoming price m (set at 0.0). The β-strategy, on the other hand, suggests more socially oriented behavior. β-investors buy when they expect an upward price trend (whenever the current price p is above a given current fundamental value v) and sell when they expect a downward price trend (whenever p is below v).

The extent to which investors follow an α-strategy or a β-strategy is weighted by the parameter S_i that represents the social susceptibility of an investor i. Stock markets and stocks alike may differ to the extent that investors focus more on fundamental characteristics of a share like price/earnings ratios and beta's, or focus more on social aspects of a share like information about which shares friends, colleagues or prominent finance experts buy. Investors may change their S given the circumstances, which leads to dynamism in the strategies they use.

The above can be formalized in the following simple formula for total excess demand:

$$E_i(p) = (1 - S_i) * (u - p) + S_i * (p - v) \tag{12.1}$$

At each time step, the price will rise when there is a positive excess demand and the price will fall when there is a negative excess demand [1]. The price is calculated as:

$$p_{t+1} = \begin{cases} E_i(p_t) > 0 \to |E_i(p_t)| + p_t * (1 - |E_i(p_t)|) \\ \text{otherwise} \to p_t * (1 - |E_i(p_t)|) \end{cases} \tag{12.2}$$

Following Day and Huang (1990), we assume α and β strategies to be individual strategies. Therefore, the excess demand is also an individual indicator of how much a single investor wants to buy or to sell. However, this leads to the problem that the total excess demand, E(p) can overpass the boundary conditions 0.0 and 1.0.

$$E(p) = \sum_{i=0}^{n} Ei(p) \tag{12.3}$$

This leads to explosive price developments and a very limited parameter space for which useful price time series can be studied. We bounded the total excess demand between 0.0 and 1.0 using an exponential transformation (12.4).

$$E(p) = 1 - \gamma \cdot \exp(-|\sum_{i=0}^{n} Ei(p)) \tag{12.4}$$

Here γ represents how strongly the market reacts to investors' actions. This parameter γ is comparable to the price adjustment coefficient c as used by Day and Huang (1990).

On the individual level, the behavior of the investors is driven by the parameter S. However, the behavior of investors is not the same in all circumstances. Investors can change their S according to their feelings and their fears (12.5). We formalize the changes in S as a combination of the agent's confidence coming from previous returns and fear coming from the deviation of the price from the fundamental value. The returns are derived from an estimation of how good individual investor agents have forecasted the price for

---

[1] This is a common way of updating the price, see e.g. Cont & Bouchaud (2000).

the next period, better forecasts implying superior returns (12.6). Investors with higher returns are expected to feel more confident. The more the current price deviates from the fundamental value, the higher the fear of investors that the stock price developments will reverse, possibly leading to losses for these investors. Therefore, at certain moments in time, trend following investors may decide to return to a more "rational" or fundamental's based strategy. This adaptation of the model (the addition of a switching mechanism in the investors' strategy) also addresses the weak point of the standard model as identified in Hoffmann et al. (2005) and more generally in Arthur (1995). This was that the market dynamics are generated by the actions of the investors, but the cognition of the investors is never affected by the evolution of the market.

$$S_i = 1 - (confidence_i * fear)_i \qquad (12.5)$$

$$confidence_i = 1 - \exp(-returns_i) \qquad (12.6)$$

$$fear_i = \exp(\frac{-(p_t - v)^2}{\delta}) \qquad (12.7)$$

$$returns_i = \frac{1}{(p_t - p_{forecasted})^2} \qquad (12.8)$$

It should be noticed, that the only parameter that is introduced in comparison to the previous version of the model is δ. This is the individual tendency of investors to be afraid. When this tendency is higher, investors will more quickly develop feelings of fear in case the current price deviates from the fundamental value. We interpret this parameter as the speed of investors' reaction to changes in the price relative to the fundamental value. We fix δ for every investors or we distribute it uniformly (e.g., $\delta=[0.0, 1.0]$). In the next section, a number of preliminary results - both *with* and *without* the switching mechanism - are discussed.

12.4 Results

In the first experiment, the influence of changing levels of trend following versus fundamental investors on the stock market dynamics is investigated.

In experiment 1.1, the level of trend following investors is uniformly distributed between 0.10 and 0.12, resulting in an average level of trend following investors of 0.11.

In experiment 1.2, the level of trend following investors is uniformly distributed between 0.1 and 0.5, resulting in an average level of trend following investors of 0.3.

In both experiments, the starting price p is 0.501, the long run investment value u and the current fundamental value v are both 0.500, and there are 100 investing agents.

The results of 500 time steps were studied [2] and it was found that with lower proportions of trend following investors (as in experiment 1.1), the standard deviation of returns is much smaller than with larger proportions of trend following investors (as in experiment 1.2). This result indicates that social interaction amongst investors may lead to an increasing level of stock market volatility, as measured by the standard deviation of the returns. This is intuitive in the sense that if an increasing number of investors rely on a social strategy to make their investment decisions, it becomes more likely that herding behavior, the corresponding stock price inflation, and increased stock market volatility occurs. Moreover, this result confirms the results of the earlier study by Hoffmann et al. (2005).

In fig. 12.1 and 12.2, the returns time series of experiments 1.1 and 1.2, respectively, are plotted.

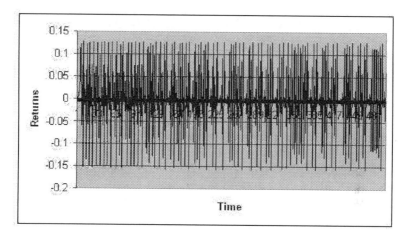

Fig. 12.1. Returns time series from experiment 1.1

In the second experiment, the influence of two different levels of the parameter δ on the stock market returns was studied. In both experiments, the initial proportion of trend following investors is 0.11, the starting price p is 0.501, the long run investment value u and the current fundamental value v are both 0.500, and there are 100 investor agents. For 500 time steps, the results were studied. In experiment 2.1, the value of δ is 0.5 and in experiment 2.2, the value of δ is 1.0.

---

[2] As a robustness check, for every simulation experiment, at least 20 runs were performed with different initial conditions.

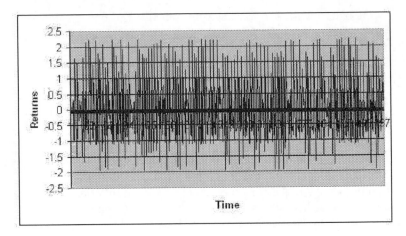

Fig. 12.2. Returns time series from experiment 1.2

It was found that when investors have a higher initial individual tendency to become afraid (indicated by a higher level of δ), the risk of previous periods becomes less important for the risk of today, as measured by ARCH [3] and GARCH [4] effects. The ARCH term represents the lagged squared error, while the GARCH term represents the lagged conditional variance. In tables 12.1 and 12.2, the ARCH and GARCH effects are displayed in the conditional variance equations for experiment 2.1 and 2.2, respectively. When investors react more fiercely to deviations of the current price from the current fundamental value, and therefore switch more easily from a trend following to a more fundamental or "rational" strategy, the stock markets become more stable, in the sense that there is less volatility clustering. So, the fear of future losses might limit the current stock market volatility.

In the third experiment, the returns for each individual agent in the agent population as aggregated over the 500 time steps of the simulation were calculated using formula refeq:hofeq8, resulting in 100 observations (one for each agent). Also, for each agent, the level of S was recorded. Scatter plots of the relationship between the level of S and the returns were made for two situations; a situation with a lower average level of S and a situation with a higher average level of S. In experiment 3.1, S was set as a uniform distribution between 0.01 and 0.21, resulting in an average level of S of 0.11. In experiment 3.2, S was set as a uniform distribution between 0.1 and 0.5, resulting in an average level of S of 0.3. This experimental design leads to the observation of the following phenomenon.

---

[3] ARCH is the test for conditional heteroscedasticity as developed by Engle(1982).

[4] GARCH is the generalized model for conditional heteroscedasticity as developed independently by Bollerslev(1986) and Taylor(1986).

Table 12.1. Conditional variance equation of experiment 2.1

	Coefficient	Std. Error	z-Statistic	Prob.
C	-0.017435	0.024994	-0.697575	0.4854
Variance Equation				
C	0.073749	0.012858	5.735721	0.0000
RESID(-1)^2	-0.136801	0.007226	-18.93076	0.0000
GARCH(-1)	0.760946	0.068734	11.07094	0.0000
R-squared	-0.001595	Durbin-Watson stat		2.355628

Table 12.2. Conditional variance equation of experiment 2.2

	Coefficient	Std. Error	z-Statistic	Prob.
C	-0.002547	0.025472	-0.099991	0.9204
Variance Equation				
C	0.157537	0.048608	3.240967	0.0012
RESID(-1)^2	-0.180182	0.012961	-13.90142	0.0000
GARCH(-1)	0.495622	0.202700	2.445101	0.0145
R-squared	-0.000020	Durbin-Watson stat		2.156982

In stock markets that are dominated by "rational" investors using a fundamental strategy as in experiment 3.1, investors with higher levels of S have higher returns than investors with lower levels of S. So, in these markets it is beneficial to be a trend following investor, and these investors can be said to be "free-riding" on the fundamental investors. In figure 12.3, this relationship is plotted.

However, in markets with a higher average level of trend following investors, as in experiment 3.2, a more complex pattern emerges. In these markets, the relationship between the level of S and the individual returns follows a U-shape, as can be seen in figure 12.4. Investors with relatively low levels of S have high returns, and so do investors with relatively high levels of S. Investors with intermediate levels of S are proverbially "stuck in the middle", as they earn lower returns than these other two groups of investors. So, in this market an investor should either be a pronounced "rational" investor following a fundamental strategy or a pronounced trend following investor in order to obtain high returns. Overall, the returns are higher in experiment 3.2 than in experiment 3.1.

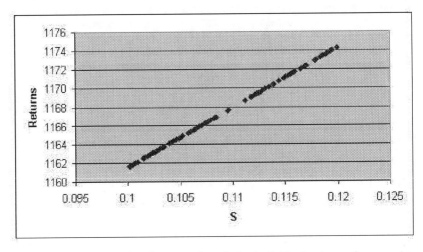

Fig. 12.3. The relationship between S and the individual returns for experiment 3.1

12.5 Conclusions

In this paper, it was shown how a relatively simple simulation model with simple micro level agent rules is capable of generating complex macro level outcomes. These outcomes of the simulation model, in the form of the returns time series, show a number of stylized facts that can also be observed in real returns time series. So, there is a qualitative resemblance between the model and the reality. However, due to e.g., the oversimplified nature of the simulation model, a quantitative gap between the results of the model and real returns time series remains.

In order to tighten or close this gap, it is necessary to radically rethink and restructure the current model in specific and the way artificial stock markets are built in general. This rethinking and restructuring may take the form of the research approach as it will be presented in one of our articles that is currently in preparation (Hoffmann, Jager & Von Eije 2006).

In general, this approach consists of four critical steps, that together constitute a complete empirical circle. Micro level agent rules are formalized based on empirical research, social interactions amongst micro level investor agents lead to macro level simulation results, macro level simulation results are subsequently compared to macro level real stock market results, and eventually

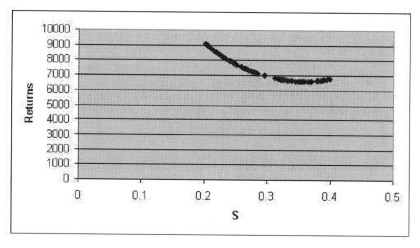

Fig. 12.4. The relationship between S and the individual returns for experiment 3.2

the simulation model can be adapted according to the results of this comparison. The final objective is to build a level 3 model of a stock market as defined by Axtell and Epstein (1994).

References

[1] Arthur W B (1995) Complexity in Economic and Financial Markets. Journal of Complexity 1
[2] Axtell R L, Epstein J M (1994) Agent-Based Models: Understanding Our Creations.
[3] Bollerslev T (1986) Generalised Autoregressive Conditional Heteroscedasticity. Journal of Econometrics 31: 307-327
[4] Cont R (2001) Empirical properties of asset returns: stylized facts and statistical issues. Quantitative Finance 1: 223-236
[5] Cont R, Bouchaud J (2000) Herd behavior and aggregate fluctuations in financial markets. Macroeconomic dynamics 4: 170-196
[6] Day R H, Huang W (1990) Bulls, bears and market sheep. Journal of Economic Behavior and Organization 14: 299-329
[7] Engle R F (1982) Autoregressive Conditional Heteroskedasticity with Estimates of the Variance of United Kingdom Inflation. Econometrica 50: 987-1007
[8] Hoffmann A O I, Delre S A, Von Eije J H, Jager W (2005) Stock Price Dynamics in Artificial Multi-Agent Stock Markets. In: P Mathieu, B Beaufils, O Brandouy (Eds.). Artificial Economics: Agent-Based Methods in Finance, Game Theory and Their Applications (pp. 191-201). Heidelberg, Springer Verlag

[9] Hoffmann A O I, Jager W (2005) The effect of different needs, decision-making processes and network structures on investor behavior and stock market dynamics: a simulation approach. ICFAI Journal of Behavioral Finance 2: 49-65

[10] LeBaron B (2005) Agent-based Computational Finance. In: K L Judd, L Tesfatsion (Eds.), The Handbook of Computational Economics Vol. II

[11] LeBaron B (2000) Agent-based computational finance: suggested readings and early research. Journal of Economic Dynamics and Control 24: 679-702

[12] Nofsinger J R (2002) The psychology of investing. Upper Saddle River, New Jersey, Prentice Hall

[13] Schleifer A (2000) Inefficient markets, an introduction to behavioral finance. Oxford University Press

[14] Shefrin H (2002) Beyond greed and fear. Understanding behavioral finance and the psychology of investing. Oxford University Press

[15] Shiller R J (2005) Irrational Exuberance. (2 ed.) New Jersey, Princeton University Press

[16] Taylor S J (1986) Forecasting the Volatility of Currency Exchange Rates. International Journal of Forecasting 3: 159-170

13

Market Polarization in Presence of Individual Choice Volatility

Sitabhra Sinha[1] and Srinivas Raghavendra[2]

[1] The Institute of Mathematical Sciences, CIT Campus, Taramani, Chennai 600113, India sitabhra@imsc.res.in

[2] Department of Economics, National University of Ireland, Galway, Ireland s.raghav@nuigalway.ie

Summary. Financial markets are subject to long periods of polarized behavior, such as bull-market or bear-market phases, in which the vast majority of market participants seem to almost exclusively choose one action (between buying or selling) over the other. From the point of view of conventional economic theory, such events are thought to reflect the arrival of "external news" that justifies the observed behavior. However, empirical observations of the events leading up to such market phases, as well events occurring during the lifetime of such a phase, have often failed to find significant correlation between news from outside the market and the behavior of the agents comprising the market. In this paper, we explore the alternative hypothesis that the occurrence of such market polarizations are due to interactions amongst the agents in the market, and not due to any influence external to it. In particular, we present a model where the market (i.e., the aggregate behavior of all the agents) is observed to become polarized even though individual agents regularly change their actions (buy or sell) on a time-scale much shorter than that of the market polarization phase.

13.1 Introduction

The past decade has seen an influx of ideas and techniques from physics into economics and other social sciences, prompting some to dub this new interdisciplinary venture as "econophysics" [1]. However, it is not just physicists who have migrated to working on problems in such non-traditional areas; social scientists have also started to use tools from, e.g., statistical mechanics, for understanding various socioeconomic phenomena as the outcomes of interactions between *agents*, which may represent individuals, firms or nations (see for example, Ref. [2]). The behavior of financial markets, in particular, has become a focus of this kind of multidisciplinary research, partly because of the large amount of empirical data available for such systems. This makes it possible to construct quantitatively predictive theories for such systems, and their subsequent validation.

Analysis of the empirical data from different financial markets has led to the discovery of several *stylized facts*, i.e., features that are relatively invariant with respect to the particular market under study. For example, it seems to be the case that markets (regardless of their stage of development) show much stronger fluctuations than would be expected from a purely Gaussian process [3, 4]. Another phenomenon that has been widely reported in financial markets is the existence of *polarized* phases, when the majority of market participants seem to opt exclusively to buy rather than sell (or vice versa) for prolonged periods. Such bull-market (or bear-market) phases, when the market exhibits excess demand (or supply) relative to the market *equilibrium* state, where the demand and supply are assumed to balance each other, are quite common and may be of substantial duration. Such events are less spectacular than episodes of speculative bubbles and crashes [5], which occur over a relatively faster time-scale; however, their impact on the general economic development of nations maybe quite significant, partly because of their prolonged nature. Hence, it is important to understand the reasons for occurrence of such market polarizations.

Conventional economic theory seeks to explain such events as reflections of news external to the market. If it is indeed true that particular episodes of market polarizations can only be understood as responses to specific historical contingencies, then it should be possible to identify the significant historical events that precipitated each polarized phase. However, although *a posteriori* explanation of any particular event is always possible, there does not seem to be any general explanation for such events in terms of extra-market variables, especially one that can be used to predict future market phases.

In contrast to this preceding approach, one can view the market behavior entirely as an emergent outcome of the interactions between the agents comprising the market. While external factors may indeed influence the actions of such agents, and hence the market, they are no longer the main determinants of market dynamics, and it should be possible to observe the various "stylized facts" even in the absence of news from outside the market. In this explanatory framework, the occurrence of market polarization can be understood in terms of time evolution of the collective action of agents. It is important to note here that the individual agents are assumed to exercise their free will in choosing their particular course of action (i.e., whether to buy or sell). However, in any real-life situation, an agent's action is also determined by the information it has access to about the possible consequences of the alternative choices available to it. In a free market economy, devoid of any central coordinating authority, the personal information available to each agent may be different. Thus the emergence of market behavior, which is a reflection of the collective action of agents, can be viewed as a self-organized coordination phenomenon in a system of heterogeneous entities.

The simplest model of collective action is one where the action of each agent is completely independent of the others; in other words, agents choose from the available alternatives at random. In the case of binary choice, where only two options are available to each agent, it is easy to see that the emergence of collective action is equivalent to a random walk on a one-dimensional line, with the number of steps equal to the number of agents. Therefore, the result will be a Gaussian distribution, with the most probable outcome being an equal number of agents choosing each alternative. As a result, for most of the time the market will be balanced, with neither excess demand nor supply. As already mentioned, while this would indeed be expected in the idealised situation of conventional economic theory, it is contrary to observations in real life indicating strongly polarized collective behavior among agents in a market. In these cases, a significant majority of agents choose one alternative over another, resulting in the market being either in a buying or selling phase. Examples of such strong bimodal behavior has been also observed in contexts other than financial markets, e.g., in the distribution of opening gross income for movies released in theaters across the USA [6].

The polarization of collective action suggests that the agents do not choose their course of action completely independently, but are influenced by neighboring agents. In addition, their personal information may change over time as a result of the outcome of their previous choices, e.g., whether or not their choice of action agreed with that of the majority [3]. This latter effect is an example of global feedback process that we think is crucial for the polarization of the collective action of agents, and hence, the market.

In this paper, we propose a model for the dynamics of market behavior which takes into account these different effects in the decision process of an agent choosing between two alternatives (e.g., buy or sell) at any given time instant. We observe a phase transition in the market behavior from an equilibrium state to a far-from-equilibrium state characterized by either excess demand or excess supply under various conditions. However, most strikingly, we observe that the transition to polarized market states occurs when an agent learns to adjust its action according to whether or not its previous choice accorded with that of the majority. One of the striking consequences of this global feedback is that, although individual agents continue to regularly switch between the alternatives available to it, the duration of the polarized phase (during which the collective action is dominated by one of the alternatives) can become extremely long. The rest of the paper is organized as follows. In the next section, we give a detailed description of the model, followed in the subsequent section by a summary of the results. We conclude with a discussion of possible extensions of the model and implications of our results. For further details please refer to Ref. [8].

---

[3] This would be the case if, as in Keynes' "beauty contest" analogy for the stock market, agents are more interested in foreseeing how the general public will value certain investments in the immediate future, rather than the long-term probable yields of these investments based on their fundamental value [7].

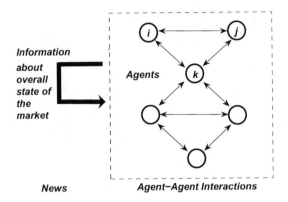

Fig. 13.1. An abstract model of a market. Each agent interacts (interactions indicated by arrows) with a subset of the other agents comprising the market, indicated by the boundary formed from the broken lines. The combined action of all agents results in the overall state of the market. The news of this state is available to all agents, although the information about the individual actions of all agents may not be accessible to any one agent.

13.2 The Model

In this section we present a general model of collective action that shows how polarization in the presence of individual choice volatility can be achieved through adaptation and learning. We assume that individual agents behave in a rational manner, where rationality is identified with actions that would result in market equilibrium in the absence of interaction between agents. Therefore, for a large ensemble of such non-interacting agents we will observe only small fluctuations about the equilibrium. Here we explore how the situation alters when agents are allowed to interact with each other. In our model, the market behavior reflects the collective action of many interacting agents, each deciding to buy or sell based on limited information available to it about the consequences of such action. An example of such limited information available to an agent is news of the overall market sentiment as reflected in market indices such as S & P 500. A schematic diagram of the various influences acting in the market is shown in Fig. 13.1.

Our model is defined as follows. Consider a population of N agents, whose actions are subject to bounded rationality, i.e., they either buy or sell an asset based on information about the action of their neighboring agents and how successful their previous actions were. The fundamental value of the asset is assumed to be unchanged throughout the period. In addition, the agents are assumed to have limited resources, so that they cannot continue to buy or sell indefinitely. However, instead of introducing explicit budget constraints [9], we have implemented gradually diminishing returns for a decision that is taken repeatedly. This is akin to the belief adaptation process in the Weisbuch-Stauffer model of social percolation [10], where making similar choices in successive periods decreases the probability of making the same choice in the subsequent period.

At any given time t, the state of an agent i is fully described by two variables: its choice, S_i^t, and its belief about the outcome of the choice, θ_i^t. The choice can be either *buy* ($= +1$) or *sell* ($= -1$), while the belief can vary continuously over a range (initially, it is chosen from a uniform random distribution). At each time step, every agent considers the average choice of its neighbors at the previous instant, and if this exceeds its belief, then it makes the same choice; otherwise, it makes the opposite choice. Then, for the i-th agent, the choice dynamics is described by:

$$S_i^{t+1} = \text{sign}(\Sigma_{j \in \mathcal{N}} J_{ij} S_j^t - \theta_i^t),\tag{13.1}$$

where $\mathcal{N}$ is the set of neighbors of agent i ($i = 1, \ldots, N$), and sign $(z) = +1$, if $z > 0$, and $= -1$, otherwise. The degree of interaction among neighboring agents, J_{ij}, is assumed to be a constant ($= 1$) for simplicity and normalized by z ($= |\mathcal{N}|$), the number of neighbors. In a lattice, $\mathcal{N}$ is the set of spatial nearest neighbors and z is the coordination number, while in the mean field approximation, $\mathcal{N}$ is the set of all other agents in the system and $z = N - 1$.

The individual belief, θ evolves over time as:

$$\theta_i^{t+1} = \begin{cases} \theta_i^t + \mu S_i^{t+1} + \lambda S_i^t, & \text{if } S_i^t \neq \text{sign}(M^t), \\ \theta_i^t + \mu S_i^{t+1}, & \text{otherwise,} \end{cases}\tag{13.2}$$

where $M^t = (1/N)\Sigma_j S_j^t$ is the fractional excess demand and describes the overall state of the market at any given time t. The adaptation rate μ governs the time-scale of diminishing returns, over which the agent switches from one choice to another in the absence of any interactions between agents. The learning rate λ controls the process by which an agent's belief is modified when its action does not agree with that of the majority at the previous instant. As mentioned earlier, the desirability of a particular choice is assumed to be related to the fraction of the community choosing it. Hence, at any given time, every agent is trying to coordinate its choice with that of the majority. Note that, for $\mu = 0, \lambda = 0$, the model reduces to the well-known zero-temperature, random field Ising model (RFIM) of statistical physics.

We have also considered a 3-state model, where, in addition to ± 1, S_i^t has a third state, 0, which corresponds to the agent choosing neither to buy nor sell. The corresponding choice dynamics, Eq. (13.1), is suitably modified by introducing a threshold, with the choice variable taking a finite value only if the magnitude of the difference between the average choice of its neighbors and its belief exceeds this threshold. This is possibly a more realistic model of markets where an agent may choose not to trade, rather than making a choice only between buying or selling. However, as the results are qualitatively almost identical to the 2-state model introduced before, in the following section we shall confine our discussion to the latter model only.

13.3 Results

In this section, we report the main results of the 2-state model introduced in the preceding section. As the connection topology of the contact network of agents is not known, we consider both the case where the agents are connected to each other at random, as well as, the case where agents are connected only to agents who are located at spatially neighboring locations. Both situations are idealised, and in reality is likely to be somewhere in between. However, it is significant that in both of these very different situations we observe market polarization phases which are of much longer duration compared to the timescale at which the individual agents switch their choice state (S).

13.3.1 Random Network of Agents and the Mean Field Model

We choose the z neighbors of an agent at random from the $N-1$ other agents in the system. We also assume this randomness to be "annealed", i.e., the next time the same agent interacts with z other agents, they are chosen at random anew. Thus, by ignoring spatial correlations, a mean field approximation is achieved.

For $z = N-1$, i.e., when every agent has the information about the entire system, it is easy to see that, in the absence of learning ($\lambda = 0$), the collective decision M follows the evolution equation rule:

$$M^{t+1} = \text{sign}[(1-\mu)M^t - \mu \Sigma_{\tau=1}^{t-1} M^\tau]. \qquad (13.3)$$

For $0 < \mu < 1$, the system alternates between the states $M = \pm 1$ (i.e., every agent is a buyer, or every agent is a seller) with a period $\sim 4/\mu$. The residence time at any one state ($\sim 2/\mu$) increases with decreasing μ, and for $\mu = 0$, the system remains fixed at one of the states corresponding to $M = \pm 1$, as expected from RFIM results. At $\mu = 1$, the system remains in the market equilibrium state (i.e., $M = 0$). Therefore, we see a transition from a bimodal distribution of the fractional excess demand, M, with peaks at non-zero values, to an unimodal distribution of M centered about 0, at $\mu_c = 1$. When we introduce learning, so that $\lambda > 0$, the agents try to coordinate with each other and at the limit $\lambda \to \infty$ it is easy to see that $S_i = \text{sign}(M)$ for all i, so that all the agents make identical choice. In the simulations, we note that the bimodal distribution is recovered for $\mu = 1$ when $\lambda \geq 1$.

For finite values of z, the population is no longer "well-mixed" and the mean-field approximation becomes less accurate the lower z is. For $z << N$, the critical value of μ at which the transition from a bimodal to a unimodal distribution occurs in the absence of learning, $\mu_c < 1$. For example, $\mu_c = 0$

for $z = 2$, while it is 3/4 for $z = 4$. As z increases, μ_c quickly converges to the mean-field value, $\mu_c = 1$. On introducing learning ($\lambda > 0$) for $\mu > \mu_c$, we again notice a transition to a state corresponding to all agents being buyers (or all agents being sellers), with more and more agents coordinating their choice.

13.3.2 Agents on a Spatial Lattice

To implement the model when the neighbors are spatially related, we consider d-dimensional lattices ($d = 1, 2, 3$) and study the dynamics numerically. We report results obtained in systems with absorbing boundary conditions; using periodic boundary conditions leads to minor changes but the overall qualitative results remain the same.

In the absence of learning ($\lambda = 0$), starting from an initial random distribution of choices and beliefs, we observe only very small clusters of similar choice behavior and the fractional excess demand, M, fluctuates around 0. In other words, at any given time an equal number of agents (on average) make opposite choices so that the demand and supply are balanced. In fact, the most stable state under this condition is one where neighboring agents in the lattice make opposite choices. This manifests itself as a checkerboard pattern in simulations carried out in one- and two-dimensional square lattices (see e.g., Fig. 13.2, top left). Introduction of learning in the model ($\lambda > 0$) gives rise to significant clustering among the choice of neighboring agents (Fig. 13.2), as well as, a large non-zero value for the fractional excess demand, M. We find that the probability distribution of M evolves from a single peak at 0, to a bimodal distribution (having two peaks at finite values of M, symmetrically located about 0) as λ increases from 0 [11]. The fractional excess demand switches periodically from a positive value to a negative value having an average residence time which increases sharply with λ and with N (Fig. 13.3). For instance, when λ is very high relative to μ, we see that M gets locked into one of two states (depending on the initial condition), corresponding to the majority preferring either one or the other choice. This is reminiscent of *lock-in* in certain economic systems subject to positive feedback [12]. The special case of $\mu = 0, \lambda > 0$ also results in a lock-in of the fractional excess demand, with the time required to get to this state increasing rapidly as $\lambda \to 0$. For $\mu > \lambda > 0$, large clusters of agents with identical choice are observed to form and dissipate throughout the lattice. After sufficiently long times, we observe the emergence of structured patterns having the symmetry of the underlying lattice, with the behavior of agents belonging to a particular structure being highly correlated. Note that these patterns are dynamic, being essentially concentric waves that emerge at the center and travel to the boundary of the region, which continually expands until it meets another such pattern. Where two patterns meet their progress is arrested and their common boundary resembles a dislocation line. In the asymptotic limit, several such patterns fill up the entire system. Ordered patterns have previously been observed in spa-

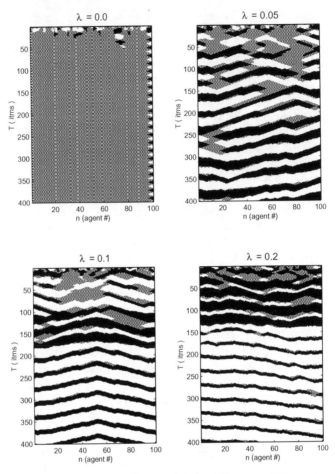

Fig. 13.2. The spatiotemporal evolution of choice (S) among 100 agents, arranged in a one-dimensional lattice, with the time-evolution upto 400 iterations starting from a random configuration shown along the vertical axis. The colors (white or black) represent the different choice states (buy or sell) of individual agents. The adaptation rate $\mu = 0.1$, and the learning rate λ increases from 0 (top left) to 0.2 (bottom right). Note that, as λ increases, one of the two states becomes dominant with the majority of agents at any given time always belonging to this state, although each agent regularly switches between the two states.

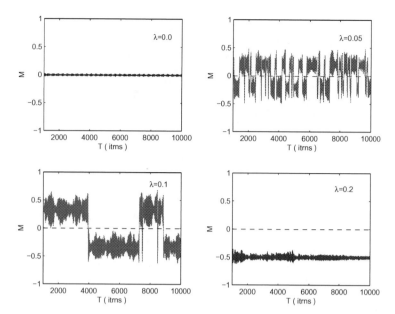

Fig. 13.3. Time series of the fractional excess demand M in a two-dimensional square lattice of 100×100 agents. The adaptation rate $\mu = 0.1$, and the learning rate λ is increased from 0 to 0.2 to show the divergence of the residence time of the system in polarized configurations.

tial prisoner's dilemma model [13]. However, in the present case, the patterns indicate the growth of clusters with strictly correlated choice behavior. The central site in these clusters act as the "opinion leader" for the entire group. This can be seen as analogous to the formation of "cultural groups" with shared beliefs [14]. It is of interest to note that distributing λ from a random distribution among the agents disrupt the symmetry of the patterns, but we still observe patterns of correlated choice behavior (Fig. 13.4). It is the global feedback ($\lambda \neq 0$) which determines the formation of large connected regions of agents having similar choice behavior.

To get a better idea about the distribution of the magnitude of fractional excess demand, we have looked at the rank-ordered plot of M, i.e., the curve obtained by putting the highest value of M in position 1, the second highest value of M in position 2, and so on. As explained in Ref. [15], this plot is related to the cumulative distribution function of M. The rank-ordering of M shows that with $\lambda = 0$, the distribution varies smoothly over a large range, while for $\lambda > 0$, the largest values are close to each other, and then shows a sudden decrease. In other words, the presence of global feedback results in a high frequency of market events where the choice of a large number of agents become coordinated, resulting in excess demand or supply. Random distribution of λ among the agents results in only small changes to the curve

186 Sitabhra Sinha and Srinivas Raghavendra

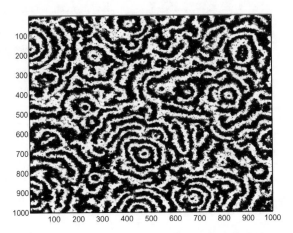

Fig. 13.4. The spatial pattern of choice (S) in a two-dimensional square lattice of 100×100 agents after 2×10^4 iterations starting from a random configuration. The adaptation rate $\mu = 0.1$, and the learning rate λ of each agent is randomly chosen from an uniform distribution between 0 and 0.1.

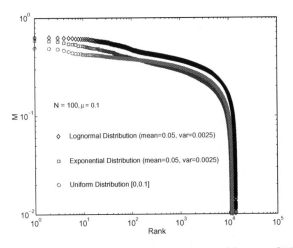

Fig. 13.5. Rank-ordered plot of M for a one-dimensional lattice of 100 agents. The adaptation rate $\mu = 0.1$, and the learning rate λ of each agent is chosen from three different random distributions: uniform (circle), exponential (square) and log-normal (diamond).

(Fig. 13.5). However, the choice of certain distribution functions for λ elevates the highest values of M beyond the trend of the curve, which reproduces an empirically observed feature in many popularity distributions that has sometimes been referred to as the "king effect" [16, 17].

13.4 Conclusion

In summary, we have presented here a model for the emergence of collective action defining market behavior through interactions between agents who make decisions based on personal information that change over time through adaptation and learning. We find that introducing these effects produces market behavior marked by two phases: (a) market equilibrium, where the buyers and sellers (and hence, demand and supply) are balanced, and (b) market polarization, where either the buyers or the sellers dominate (resulting in excess demand or excess supply). There are multiple mechanisms by which the transition to market polarization occurs, e.g., (i) keeping the adaptation and learning rate fixed but switching from an initially regular neighborhood structure (lattice) to a random structure (mean-field) one sees a transition from market equilibrium to market polarization; (ii) in the lattice, by increasing the learning rate λ (keeping μ fixed) one sees a transition from equilibrium to polarization behavior; and (iii) in the case where agents have randomly chosen neighbors, by increasing the adaptation rate μ beyond a critical value (keeping λ fixed) one sees a transition from polarized to equilibrium market state.

The principal interesting observation seems to be that while, on the one hand, individual agents regularly switch between alternate choices as a result of adapting their beliefs in response to new information, on the other hand, their collective action (and hence, the market) may remain polarized in any one state for a prolonged period. Apart from financial markets, such phenomena has been observed, for example, in voter behavior, where preferences have been observed to change at the individual level which is not reflected in the collective level, so that the same party remains in power for extended periods. Similar behavior possibly underlies the emergence of cooperative behavior in societies. As in our model, each agent can switch regularly between cooperation and defection; however, society as a whole can get trapped in a non-cooperative mode (or a cooperative mode) if there is a strong global feedback.

Even with randomly distributed λ we see qualitatively similar results, which underlines their robustness. In contrast to many current models, we have not assumed a priori existence of contrarian and trend-follower strategies among the agents [18]. Rather, such behavior emerges naturally from the micro-dynamics of agents' choice behavior. Further, we have not considered external information shocks, so that all observed fluctuations in market activity is endogenous. This is supported by recent empirical studies which have failed to observe any significant correlation between market movements and exogenous economic variables like investment climate [19].

We have recently studied a variant of the model in which the degree of interactions between neighboring agents J_{ij} is not uniform and static, but evolves in time [20]. This is implemented by assuming that agents seek out the most successful agents in its neighborhood, and choose to be influenced

188 Sitabhra Sinha and Srinivas Raghavendra

by them preferentially. Here, *success* is measured by the fraction of time the agents decision (to buy or sell) accorded with the market behavior. The resulting model exhibits extremely large fluctuations around the market equilibrium state ($M = 0$) that quantitatively match the fluctuation distribution of stock price (the "inverse cubic law") seen in real markets.

Another possible extension of the model involves introducing stochasticity in the dynamics. In real life, the information an agent obtains about the choice behavior of other agents is not completely reliable. This can be incorporated in the model by making the updating rule Eq. (13.1) probabilistic. The degree of randomness can be controlled by a "temperature" parameter, which represents the degree of reliability an agent attaches to the information available to it. Preliminary results indicate that higher temperature produces unimodal distribution for the fractional excess demand.

Our results concerning the disparity between behavior at the level of the individual agent, and that of a large group of such agents, has ramifications beyond the immediate context of financial markets [21]. As for example, it is often said that "democracies rarely go to war" because getting a consensus about such a momentous event is difficult in a society where everyone's free opinion counts. This would indeed have been the case had it been true that the decision of each agent is made independently of others, and is based upon all evidence available to it. However, such an argument underestimates how much people are swayed by the collective opinion of those around them, in addition to being aroused by demagoguery and yellow journalism. Studying the harmless example of how market polarizations occur even though individuals may regularly alternate between different choices may help us in understanding how more dangerous mass madness-es can occur in a society.

Acknowledgements

We thank J. Barkley Rosser, Bikas Chakrabarti, Deepak Dhar, Matteo Marsili, Mishael Milakovic, Ram Ramaswamy, Purusattam Ray and Dietrich Stauffer for helpful discussions. SS would like to thank the Santa Fe Institute where part of the work was done and Sam Bowles, Jung-Kyoo Choi, Doyne Farmer and Lee Segel for comments.

References

[1] Farmer J D, Shubik M, Smith E (2005) Is economics the next physical science ? Physics Today 58 (9):37–42
[2] Durlauf S N (1999) How can statistical mechanics contribute to social science ? Proc. Natl. Acad. Sci. USA 96: 10582–10584
[3] Gopikrishnan P, Meyer M, Amaral L A N, Stanley H E (1998) Inverse cubic law for the distribution of stock price variations, Eur. Phys. J. B 3:139140

[4] Sinha S, Pan R K (2006) The power (law) of Indian markets: Analysing NSE and BSE trading statistics. In: Chatterjee A, Chakrabarti B K (ed) Econophysics of stock and other markets. Springer, Milan

[5] Schiller R J (2000) Irrational exuberance. Princeton University Press, Princeton

[6] Sinha S, Raghavendra S (2004) Hollywood blockbusters and long-tailed distributions: An empirical study of the popularity of movies. Eur. Phys. J. B 42: 293–296

[7] Keynes J M (1934) The general theory of employment, interest and money. Harcourt, New York

[8] Sinha S, Raghavendra S (2004) Phase transition and pattern formation in a model of collective choice dynamics. SFI Working Paper 04-09-028

[9] Iori G (2002) A microsimulation of traders activity in the stock market: the role of heterogeneity, agents interaction and trade frictions, J. Economic Behavior & Organization 49:269285

[10] Weisbuch G, Stauffer D (2003) Adjustment and social choice. Physica A 323: 651–662

[11] Sinha S, Raghavendra S (2005) Emergence of two-phase behavior in markets through interaction and learning in agents with bounded rationality. In: Takayasu H (ed) Practical fruits of econophysics. Springer, Tokyo :200-204

[12] Arthur B W (1989) Competing technologies, increasing returns, and lock-in by historical events. Economic J. 99: 116–131

[13] Nowak M A, May R M (1992) Evolutionary games and spatial chaos. Nature 359: 826–829

[14] Axelrod R (1997) The dissemination of culture: A model with local convergence and global polarization. J. Conflict Resolution 41: 203–226

[15] Adamic L A, Huberman B A (2002) Zipf's law and the internet. Glottometrics 3:143–150

[16] Laherrere J, Sornette D (1998) Stretched exponential distributions in nature and economy: "fat tails" with characteristic scales. Eur. Phys. J. B 2: 525–539

[17] Davies J A (2002) The individual success of musicians, like that of physicists, follows a stretched exponential distribution. Eur. Phys. J. B 4: 445–447

[18] Lux T (1995) Herd behaviour, bubbles and crashes. Economic J. 105: 881–896

[19] Kaizoji T (2000) Speculative bubbles and crashes in stock markets: An interacting-agent model of speculative activity. Physica A 287: 493–506

[20] Sinha S (2006) Apparent madness of crowds: Irrational collective behavior emerging from interactions among rational agents. In: Chatterjee A, Chakrabarti B K (ed) Econophysics of stock and other markets. Springer, Milan

[21] Sinha S, Pan R K (2006) How a "hit" is born: The emergence of popularity from the dynamics of collective choice. In: Chatterjee A, Chakraborti

A, Chakrabarti B K (eds) Handbook of econophysics and sociophysics, Wiley-VCH

Is Ignoring Public Information Best Policy? Reinforcement Learning in Information Cascade

Toshiji Kawagoe and Shinichi Sasaki

Future University - Hakodate, 116-2 Kameda Nakano cho, Hakodate, Hokkaido, 041-8655 Japan, kawagoe@fun.ac.jp

14.1 Introduction

In this paper, we examine an economic theory of information cascade[1], which predicts a type of herding behaviors, in an agent-based simulation.

The economic theory of information cascade tries to explain herding behaviors of agents by a statistical decision model. In that model, each agent decides sequentially one by one. Each agent can observes both a noisy signal of the state of the world (e.g., quality of a product) and the other agents' behaviors (e.g., purchasing decisions) which were made previously. The theory stresses importance of public information of the past histories of the other agents' choices. Surprisingly, according to that theory, an agent can be easily affected by only a few agents who decided before and tend to ignore his/her own private taste or preference, then follow the predecessors' actions. If this theory is truly applicable to the real world situation such as a customer marketing, providing public information such as purchasing histories of the other agents has in fact a great impact on improving the performance of customer marketing.

On the other hand, the theory of information cascade ignore the effect of agent's learning, namely, it considers only one-shot, static environment. As it is unrealistic assumption, we develop a new model featured by a reinforcement learning[2] to extend the model of information cascade, and conducted a series of experiments in order to check which factor, public information or agent's learning, is crucial to induce information cascade in a dynamic environment.

Our results show that incorporating the customer's past experience into the model has great impact on the emergence of information cascade.

---

[1] Birkhchandani et al. [5], Chamley [7], and Gale [10] are useful survey of the theory of information cascades.

[2] See Sutton and Barto [16] and Young [17].

The organization of the paper is as follows. In the next section, we briefly summarize the theory of information cascade. In section 14.4, the design and results of our simulation is shown. Conclusions are given in the last section.

14.2 The Theory of Information Cascade

In many socio-economic situations, although agents have private information or taste/preference of the item to be purchased, they tend to follow the decisions made by their predecessors. An information cascade occurs when it is optimal for each agent to ignore its private information and follow the decisions made by the other agents who have decided before.

The theory of information cascade was initiated by several researchers in economics such as Banerjee [3] and Birkhchandani et al. [4], and it has been tested in the laboratory with human subjects by a number of experimental economists (see, for example, Allsopp and Hey [1], Anderson and Holt [2], Kübler and Weizsäcker [11] [12], Hung and Plott [13], Sasaki [14], Sasaki and Kawagoe [15]). It is also applied for explaining herding behavior in financial decision making (Cipriani and Guarino [8] and Drehmann et al. [9]).

In the theory of information cascade, each agent makes a decision sequentially one by one. Each agent's task is to predict a realized, true state of the world, $t \in T$ (in our context, for example, quality of a product). Each agent knows prior probability distribution, $p(t)$, of the state of the world t. Before agent's decision is made, agent i receives a noisy signal or information of the state of the world, s_i, privately. Each agent also knows likelihood, $p(s_i|t)$, of receiving that noisy signal s_i conditional of the state t. In addition, publicly announced past decisions made by the other agents (public information), $\{a_j\}_{j=1}^{i-1}$, is available. So agent i can calculate postrior probability, $p(t|s_i, \{a_j\}_{j=1}^{i-1})$, of the realized state of world by utilizing its private information $s_i(t)$, likelihood $p(s_i|t)$ and public information $\{a_j\}_{j=1}^{i-1}$ to predict true state of the world.

But if a small number of agents accidentally made same decisions, then it is easily shown by Bayes rule that an agent who observes such public information should ignore its private information and follows its predecessors' decisions. In other words, for $t, t'(t \neq t')$ and $\forall s_i, s_i'(s_i \neq s_i')$, if

$$p(t|s_i, \{a_j\}_{j=1}^{i-1}) > p(t'|s_i, \{a_j\}_{j=1}^{i-1})$$

then

$$p(t|s_i', \{a_j\}_{j=1}^{i-1}) > p(t'|s_i', \{a_j\}_{j=1}^{i-1})$$

holds when information cascade occurs. Thus the theory of information cascade represents fads or herding behaviors of agents[3].

---

[3] As Çelen and Kariv [6] pointed out, information cascade defined by Birkhchandani et al. [4] and herd behavior defined by Banerjee [3] are slightly diffenrent. The

If the prediction made by agents who are involved in an information cascade is true state of the world, we call it "good cascade." Of course, it is possible that the prediction made by agents who are involved in an information cascade is not correct state of the world. We call such a case "bad cascade."

Now let us examine whether the theoretical prediction of the theory of information cascade can be verified in an agent-based simulation.

14.3 Agent-Based Simulation

In our simulation, five agents in a group predict the true state of the world sequentially. Following the basic setup of experiments conducted by Anderson and Holt [2], we compare their symmetric case with asymmetric one in our simulation. In each case, there are two state of the world, A and B, and two signals, a and b. In symmetric case, prior probability of the state A and B are $p(A) = p(B) = 1/2$ respectively, and likelihood of receiving a signal a (b) in the state A (B) are $p(a|A) = p(b|B) = 2/3$. In this case, at least two agents make same prediction, an agent who observes such public information follows the predecessors' prediction. In asymmetric case, prior probability of the state A and B are $p(A) = p(B) = 1/2$ respectively, and likelihood of receiving a signal a (b) in the state A (B) are $p(a|A) = 6/7$ and $p(b|B) = 2/7$. In this case, if at least four agents predict A, then predicting A is best response for an agent who observes such public information, and if at least an agent predict B, then predicting B is best response for subsequent agents whatever signal they receive.

We compare the following three agent models; (1) Agents who follows Bayes rule (Type B) behaves in accordance with Bayes rule given its private information and the predecessor's decisions as in theoretical prediction. This case is as a benchmark, so that agent's own private past experiences of prediction are not available. (2) Agents who follow reinforcement learning without public information (Type RNP) can see only its private information, and follows a reinforcement learning (see Sutton and Barto [16] and Young [17]). Public information is not available for each agent in this case, but agents can change their behaviors in accordance with their own private past experiences of prediction. (3) Agents who follow reinforcement learning with public information (Type RP) can see not only its private information but also public information, i.e., the decisions made by the other agents who decided before that agent. So agents can utilize not only their own private past experiences of prediction but also public information.

For Type B and RNP model, theoretical frequency of information cascade can be derived from Bayes rule. We assume that each agent declares a state

---

difference between these notions is apparent when the domain of the state of world is continuous. But we can use these terms interchangeably when the domain of the state of world is discrete.

with higher posterior probability. When the posterior probability for each state is equal, either is declared with equal probability. This is our tie-breaking rule. Let m be the number of signal a observed and n be the number of signal b observed. For $m+n+1$st agent, in symmetric case, posterior probability that true state is A, $p(A|m,n)$, is as follows.

$$p(A|m,n) = \frac{p(m,n|A)p(A)}{p(m,n|A)p(A) + p(m,n|B)p(B)}$$
$$= \frac{(2/3)^m(1/3)^n(1/2)}{(2/3)^m(1/3)^n(1/2) + (1/3)^m(2/3)^n(1/2)} = \frac{2^m}{2^m + 2^n}$$

Similarly, posterior probability that true state is A in asymmetric case is derived as follows.

$$p(A|m,n) = \frac{p(m,n|A)p(A)}{p(m,n|A)p(A) + p(m,n|B)p(B)}$$
$$= \frac{(6/7)^m(1/7)^n(1/2)}{(6/7)^m(1/7)^n(1/2) + (5/7)^m(2/7)^n(1/2)} = \frac{6^m}{6^m + 5^m 2^n}$$

Using these equations, one can derive theoretical frequency of information cascade for Type B and RNP agents. Table 14.1 and 14.2 shows that frequency of information cascade in each length. Here A-cascade means that an information cascade in which agents predict A occurs, and B-cascade means that an information cascade in which agents predict B occurs.

Table 14.1. Theoretical Frequency of Information Cascade (Symmetric case)

Length	Type	A-cascade B	A-cascade RNP	B-cascade B	B-cascade RNP
3	Good	0.062	0.230	0.062	0.230
	Bad	0.025	0.049	0.025	0.049
	Total	0.087	0.0279	0.087	0.279
4	Good	0.062	0.132	0.062	0.132
	Bad	0.025	0.016	0.025	0.016
	Total	0.087	0.148	0.087	0.148
5	Good	0.556	0.132	0.556	0.132
	Bad	0.222	0.004	0.222	0.004
	Total	0.778	0.136	0.778	0.136

We used a classifier system like reinforcement learning for Type RP and RNP models in our simulation. Given the private information s_i and public

Table 14.2. Theoretical Frequency of Information Cascade (Asymmetric case)

Length	Type	A-cascade		B-cascade	
		B	RNP	B	RNP
3	Good	0.064	0.039	0.052	0.030
	Bad	0.105	0.058	0.073	0.002
	Total	0.169	0.097	0.125	0.032
4	Good	0.000	0.154	0.102	0.074
	Bad	0.122	0.148	0.000	0.010
	Total	0.122	0.302	0.102	0.084
5	Good	0.556	0.463	0.285	0.186
	Bad	0.143	0.000	0.260	0.001
	Total	0.699	0.463	0.545	0.187

information $\{a_j\}_{j=1}^{i-1}$, each agent chooses one of the predictions, A or B, with probability $p_i^A(s_i, \{a_j\}_{j=1}^{i-1}|\tau)$ and $p_i^B(s_i, \{a_j\}_{j=1}^{i-1}|\tau)$ respectively in time period τ. p_i^A and p_i^B are determined by past experiences of the predictions made by agent i. Denote $f_i^A(s_i, \{a_j\}_{j=1}^{i-1}|\tau)$ and $f_i^B(s_i, \{a_j\}_{j=1}^{i-1}|\tau)$ as the "fitness" of predicting A and B in time period τ given s_i and $\{a_j\}_{j=1}^{i-1}$. $f_i^A(s_i, \{a_j\}_{j=1}^{i-1}|\tau)$ and $f_i^B(s_i, \{a_j\}_{j=1}^{i-1}|\tau)$ are updated as follows. First we assume

$$f_i^A(s_i, \{a_j\}_{j=1}^{i-1}|0) = f_i^B(s_i, \{a_j\}_{j=1}^{i-1}|0) = 1$$

as initial conditions. Then if agent i predicts A and it is correct (incorrect) at period τ, the fitness of the prediction A is updated as follows;

$$f_i^A(s_i, \{a_j\}_{j=1}^{i-1}|\tau+1) = \beta f_i^A(s_i, \{a_j\}_{j=1}^{i-1}|\tau) \pm R$$

where $R > 0$ is a constant for correct (incorrect) prediction and $\beta(0 < \beta \leq 1)$ is forgetting parameter. If agent i did not predict A at period τ, the fitness of the predicting A is updated as follows;

$$f_i^A(s_i, \{a_j\}_{j=1}^{i-1}|\tau+1) = \beta f_i^A(s_i, \{a_j\}_{j=1}^{i-1}|\tau).$$

Then, given the fitness of the prediction, probability of predicting A and B at time period τ is given in the following logit form,

$$p_i^A(s_i, \{a_j\}_{j=1}^{i-1}|\tau) = \frac{\exp(f_i^A(s_i, \{a_j\}_{j=1}^{i-1}|\tau))}{\exp(f_i^A(s_i, \{a_j\}_{j=1}^{i-1}|\tau)) + \exp(f_i^B(s_i, \{a_j\}_{j=1}^{i-1})|\tau)}$$

The updating rule and probability of predicting B is defined analogously. Totally 1000 rounds of simulation are conducted in each model.

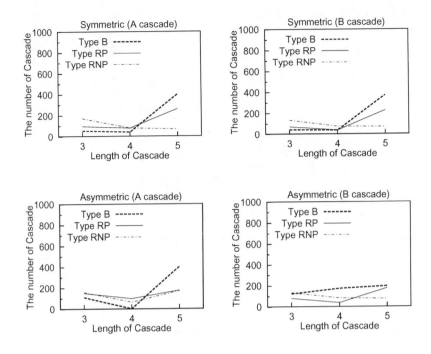

Fig. 14.1. The Number of Occurrence of Information Cascade

14.4 Results

Fig. 14.1 depicts the number of occurrence of information cascade in each condition. Information cascades occurred in our simulation are classified by their length. The length of information cascade is measured by the number of agents who take same predictions subsequently. So if three agents subsequently predict A, then we can say that information cascade with length 3 occurred.

From Fig. 14.1, one can easily see that Type B agents likely form longer cascade. In fact, information cascades with length 5 occur most frequently in Type B model in any case. On the other hand, information cascades with length 3 occur most frequently in Type RNP model in every condition. The results of Type RP model lie in between them.

Table 14.3 shows the frequency of good and bad cascades. It is quite surprising fact that, for any length, good cascades occur most frequently in Type RNP model in symmetric A-cascade cases. In addition, in symmetric B-cascade and asymmetric A-cascade cases, frequency of good cascades with length 4 and 5 are also the highest in Type RNP model. On the other hand, Type B model outperforms other models for predicting good cascades only in symmetric B-cascade with length 3 and asymmetric B-cascade cases.

Table 14.3. The Proportion of Occurrences of Good Cascade

Symmetric case

Length	Type	A-cascade B	RP	RNP	B-cascade B	RP	RNP
3	Good	39 (75.0%)	59 (62.1%)	132 (78.1%)	33 (78.6%)	46 (67.6%)	102 (77.9%)
	Bad	13 (25.0%)	36 (37.9%)	37 (21.9%)	9 (21.4%)	22 (32.4%)	29 (22.1%)
	Total	52 (100%)	95 (100%)	169 (100%)	42 (100%)	68 (100%)	131 (100%)
4	Good	28 (70.0%)	52 (66.6%)	63 (79.7%)	32 (84.2%)	28 (75.7%)	62 (84.9%)
	Bad	12 (30.0%)	26 (33.4%)	16 (20.3%)	6 (15.8%)	9 (24.3%)	11 (15.1%)
	Total	40 (100%)	78 (100%)	79 (100%)	38 (100%)	37 (100%)	73 (100%)
5	Good	299 (74.5%)	199 (75.9%)	61 (91.0%)	264 (70.9%)	175 (77.4%)	65 (95.5%)
	Bad	102 (25.5%)	63 (24.1%)	6 (8.0%)	108 (29.1%)	51 (22.6%)	3 (4.5%)
	Total	401 (100%)	262 (100%)	67 (100%)	372 (100%)	226 (100%)	68 (100%)

Asymmetric case

Length	Type	A-cascade B	RP	RNP	B-cascade B	RP	RNP
3	Good	35 (33.3%)	66 (42.0%)	57 (38.5%)	74 (60.7%)	76 (58.0%)	41 (52.6%)
	Bad	74 (66.7%)	91 (58.0%)	91 (61.5%)	48 (39.3%)	55 (42.0%)	37 (47.4%)
	Total	109 (100%)	157 (100%)	148 (100%)	122 (100%)	131 (100%)	78 (100%)
4	Good	0 (0.0%)	29 (46.0%)	49 (50.5%)	110 (100.0%)	48 (59.3%)	25 (67.6%)
	Bad	63 (100.0%)	34 (54.0%)	48 (49.5%)	0 (0.0%)	33 (40.7%)	12 (32.4%)
	Total	63 (100%)	63 (100%)	97 (100%)	110 (100%)	81 (100%)	37 (100%)
5	Good	266 (66.5%)	118 (69.0%)	118 (67.4%)	137 (69.9%)	39 (52.7%)	99 (55.9%)
	Bad	134 (33.5%)	53 (31.0%)	57 (32.6%)	59 (30.1%)	35 (47.3%)	78 (44.1%)
	Total	400 (100%)	171 (100%)	175 (100%)	196 (100%)	74 (100%)	177 (100%)

This means that the prediction made by the agents who utilizes only their own private past experiences of prediction is more likely to be true except for asymmetric B-cascade cases. So, ignoring public information, and believing one's own private information and experiences is much better for making correct prediction.

14.5 Conclusions

After the first well-controlled laboratory experiment by Anderson and Holt [2] confirmed theoretical prediction of information cascade made by Birkhchandani et al. [4], their result has been replicated by an experimentalist such as Hung and Plott [13]. But a number of experimentalists such as Kübler and Weizsäcker [11] [12], and Sasaki and Kawagoe [15] pointed out that theoretical prediction of Birkhchandani et al. [4] does not hold in exact sense in the laboratory. They showed that subjects' behaviors were still affected by their private signals even after a longer cascade has already occurred. Then, from such results, a natural question arises in our mind whether ignoring public information is best policy for predicting true state of world.

In this paper, we tried to answer that question by developing a new model featured by a reinforcement learning to extend the model of information cascade, and conducted a series of experiments in order to check which factor, public information or agent's learning, is crucial to induce information cascade in a dynamic environment.

Our results showed that incorporating agent's past experience of prediction had great impact on the emergence of information cascade and that providing public information was rather detrimental for agents who would like to predict correct state of the world. Thus, the prediction made by the agents who utilizes only their own private past experiences of prediction is more likely to be true except for asymmetric B-cascade cases. So, ignoring public information, and believing one's own private information and experiences is much better for making correct prediction. This confirm the tendency of subjects utilizing their private information even after information cascade started, which was observed by Kübler and Weizsäcker [11] [12], and Sasaki and Kawagoe [15].

References

[1] Allsopp, L. and J. D. Hey: Two Experiments To Test A Model Of Herd Behavior, Experimental Economics, 3 (2000) 121-136
[2] Anderson, L. R. and C. A. Holt: Information Cascades in the Laboratory, American Economic Review, 87 (1997) 847-862
[3] Banerjee, A. V.: A Simple Model of Herd Behavior, Quarterly Journal of Economics, 57 (1992) 797-817

[4] Birkhchandani, S., D. Hirshleifer, and I. Welch: A Theory of Fads, Fashion, Custom, and Cultural Change as Informational Cascades, Journal of Political Economy, 100 (1992) 992-1026

[5] Birkhchandani, S., D. Hirshleifer, and I. Welch: Learning From the Behavior of Others: Conformity, Fads, and Informational Cascades, Journal of Economic Perspectives, 12 (1998) 151-170

[6] Çelen, B., S. Kariv: Distinguishing Information Cascades From Herd Behavior In The Laboratory, American Economic Review, 94 (2004) 484-498

[7] Chamley, C. P.: Rational Herds Economic Models Of Social Learning, Cambridge University Press, 2004

[8] Cipriani, M., A. Guarino: Herd Behavior In A Laboratory Financial Market, American Economic Review, 95 (2005) 1427-1443

[9] Drehmann, M., J. Oechssler, A. Roider: Herding And Contrarian Behavior In Financial Markets: An Internet Experiment, American Economic Review, 95 (2005) 1403-1426

[10] Gale, D.: What Have We Learned From Social Learning? European Economic Review, 40 (1996) 617-628

[11] Kübler, D. and G. Weizsäcker: Limited Depth Of Reasoning And Failure Of Cascade Formation In The Laboratory, Review of Economic Studies, 71 (2004) 425-441

[12] Kübler, D. and G. Weizsäcker: Are Longer Cascades More Stable? mimeo

[13] Hung, A. A., C. R. Plott: Information Cascades: Replication And An Extension To Majority Rule And Conformity-Rewarding Institutions, American Economic Review, 91 (2001) 1508-1520

[14] Sasaki, S.: Signal Qualities, Order Of Decisions, And Informational Cascades: Experimental Evidence, Economics Bulletin, 3 (2005) 1-11

[15] Sasaki, S. and T. Kawagoe: Can You Believe Your Neighbors' Behaviors? mimeo

[16] Sutton, R. S., Barto, A. G.: Reinforcement Learning An Introduction. The MIT Press, 1998

[17] Young, H. P.: Strategic Learning And Its Limit. Oxford University Press, 2004

Part V

Social Interaction - Connectivity

Complex Behaviours in Binary Choice Model with Global or Local Social Influence

Denis Phan and Stéphane Pajot

CREM UMR CNRS 6211, University of Rennes 1, France
denis.phan@univ-rennes1.fr

Summary. This paper illustrates the effects of global or local social influences upon binary choice. Analytical results are summarized and an ACE (Agent based Computational Economics) approach is used to investigate the corresponding mechanisms of interdependence in the case of a coordination problem and finite size effects.

15.1 Introduction

In this paper, we explore the effects of the introduction of social influences through fixed interaction structures upon local and global properties of a simple model of binary choice. More specifically, interlinked agents have to make a binary choice. Their preferences are both *intrinsically heterogeneous* (idiosyncratic preferences) and *interactively heterogeneous* (it positively depends on the choice of their neighbours). Aggregate outcomes of such situation may be characterized by multiple equilibria and complex dynamics with "tipping" and "avalanches". The first part of the present study summarizes analytical results in case of global influence while the second part relies on numerical simulations in the case of finite size population for both a global and a local influence network, making use of "Moduleco-Madkit', a multi-agent platform (Gutknecht and Ferber 2000; Phan 2004; Michel *et al.* 2005).

15.1.1 A Short Birds Eyes View of the Literature

The question of binary choices with externalities in the social sciences has been directly addressed by (Schelling 1973, 1978), and the question of individual and collective threshold of adoption has been introduced later by (Granovetter, 1978). In such models, the *individual threshold of adoption* is defined as the number of adopters each agent considers to be sufficient to modify his behaviour. As a result, the final equilibrium depends on the distribution of individual thresholds, and in numerous cases with several equilibria, the selection of a particular equilibrium depends on the history of the collective

204   Denis Phan and Stéphane Pajot

dynamics. In the context of "global influence", there is no "local network" in the sense that individuals are only sensitive to the percentage of the total population which has previously adopted (a behaviour, a good, a service etc.). (Valente 1995) stresses the importance of the local structure of interpersonal relations in the propagation phenomenon (innovations, opinions), and defines the *threshold of exposure* of an agent as the proportion of adopters in his personal network (neighbourhood) sufficient enough to induce a change in his behaviour.

In the mathematical sociology field, (Weidlich and Haag 1983) proposes, in the global perspective, a generic model of opinion formation based upon a master equation and the Fokker-Plank approximation approach. In the micro-to macro perspective,(Kindermann and Snell 1980) identifies a social network as an application of a Markov random field. (Galam *et al.* 1982) proposes probably the first micro-based application of statistical physics tools to sociology [1]. This pioneering paper proposes a new approach of tipping in collective behaviour applied to strikes. But the scope of this paper is quite larger. Galam and co-authors identify by the way of a "phase analysis" the existence of two regimes (or "phases") separated by a critical point, in the neighbourhood of which the system is extremely sensitive to small changes in parameters as well as to the history of the system. Then, by a tipping effect, small microscopic changes can lead to drastic changes at the macro level.

In economics, the pioneering work of (Föllmer 1974) considers local stochastic interactions by the way of Markov random fields in a general equilibrium model with random preferences. The same year, Gary Becker advocates the introduction of social environment and social interactions in the rational decision of individuals, through his concept of "social income" (Becker 1974). In the middle of the 80's, (Kirman 83) and (Kirman and Oddou and Weber 1986) suggests the use of stochastic graph theory in order to take into account the local communications between agents within the markets. But the real take off for the models of individual choice with interactions and social influence in economics began by the 90's. Some typical contributions are (Brock and Durlauf 2001a), (Glaeser and Sacerdote and Scheinkman, 1996; Glaeser and Scheinkman 2002) for the emphasis on social dimension in a Beckerian tradition, and (Ioannides 2006) for the topologies of interactions [2].

The model briefly discussed in this paper- hereafter referred as the GNP model - was previously presented elsewhere in (Gordon *et al.* 2005), (Nadal *et al.* 2005), (Phan and Semeshenko 2006), and generalized to a large class of distributions in (Gordon *et al.* 2006). The general structure of the GNP model seems to be reminiscent of a class of models by Durlauf and co-authors (Blume and Brock 2001) and especially (Brock and Durlauf 2001a, 2001b)

---

[1] This approach is qualified as "sociophysics". For a discussion of the relationship with mechanical physics, see (Durlauf 1999), and (Phan and Nadal and Gordon 2004).

[2] see syntheses by (Blume 1997; Durlauf 1997; Blume and Durlauf 2001b)

- hereafter referred as the DBB model. But this apparent similarity is only superficial and the structure of GNP and DBB differs by the nature of the disorder (e.g. heterogeneity across agents and randomness). Therefore, in the GNP model agents are heterogeneous with respects to their idiosyncratic preferences, which remain fixed and do not contain additively stochastic term, while the DBB model belongs to the class of both *Random Utility Model* (RUM) [3] and *quantal choice analysis* (McFadden 1974). The DBB model assumes a double exponential (extreme value, type I) independent identically distributed random variables in each sub-utility of the underlying Thurstone's *discriminal process*, hence the distribution function for the difference of these random variables is logistic. As underlined elsewhere (Phan *et al.* 2004; Nadal *et al.* 2005), these two classes of models are quite different. The DBB model belongs to the class of the Classic Ising Model with "annealed" disorder. The heterogeneity comes from the random term of the RUM only, not from the deterministic term, assumed to be the same for all agents On the contrary, our own model is formally equivalent to a "Random Field Ising Model" (RFIM), with a fixed heterogeneous idiosyncratic term: the disorder is said to be "quenched" (i.e. there is no random utility). These two kinds of models can lead to very different behaviours (Stanley 1971; Galam and Aharony 1980, 1981; Galam 1982; Sethna *et al.* 1993, 2005). Some of them are presented below.

15.2 The Model and its Global Behaviour

The question of social influence over individual choice is now on the economist's agenda. In this section, some analytical results from the GNP model are presented and discussed in the particular case of global influence and symmetric triangular distribution of idiosyncratic preferences (Phan and Semeshenko 2006).

15.2.1 Modelling the Individual Choice in a Social Context

We consider a set of N agents $i \in \Lambda_N \equiv \{1, 2, .., N\}$ with a classical linear willingness-to-adopt function. Each agent makes a simple binary choice, that is, either adopts ($\omega_i = 1$) or does not adopt ($\omega_i = 0$). A rational agent chooses ω_i in the strategic set $\Omega \equiv \{0, 1\}$ in order to maximize a linear surplus function $\omega_i V_i$:

$$W_i\left(\omega_i \mid \tilde{\omega}_{-i}\right) \equiv \max_{\omega_i \in \{0,1\}} \left\{\omega_i . V_i\left(\tilde{\omega}_{-i}\right)\right\}$$
$$\text{with}: \quad V_i\left(\tilde{\omega}_{-i}\right) = (H_i - C) + \frac{J_{ik}}{N_{\vartheta i}} \sum_{k \in \vartheta_i} \tilde{\omega}_k \qquad (15.1)$$

---

[3] Originated in Thurstone's model of comparative judgment (Thurstone 1927), introduced in economics by (Marschak 1960; Block and Marschak 1960), see also (Mansky 1977)

Where C is the cost of adoption [4] and H_i represents the idiosyncratic preference component. Some other agents k, influence agent i's preferences through their own choices ω_k. Agents k hereafter called neighbours of i are within a subset: $\vartheta_i \in \Lambda_N$, of size $N_{\vartheta i}$, called neighbourhood of i such that each agent $k \in \vartheta_i$. This social influence is represented here by a weighted sum of these choices. Let us denote $J_{ik}/N_{\vartheta i}$ the corresponding weight i.e. the marginal social influence on agent i, from the decision of agent $k \in \vartheta_i$. This social influence is assumed to be positive: $J_{ik} > 0$. For simplicity, we consider here only the case of homogeneous influences, that is, identical positive weights for all influence parameters in the neighbourhood: $\forall i \in \Lambda_N, \forall k \in \vartheta_i : J_{ik} = J$. For a given neighbour k taken in the neighbourhood ϑ_i, the marginal social influence is $J/N_{\vartheta i}$ if the neighbour is an adopter ($\omega_i = 1$), and zero otherwise. The individual surplus (15.1) can be rewritten in a more simply way as:

$$W_i\left(\omega_i \,|\, \tilde{\omega}_{-i}\right) \equiv \max_{\omega_i \in \{0,1\}} \left\{\omega_i\left(H_i - C + J\eta_i^e(\tilde{\omega}_{-i})\right)\right\}$$

with : $$\eta_i^e(\tilde{\omega}_{-i}) \equiv \sum_{k \in \vartheta_i} \tilde{\omega}_k / N_{\vartheta i} \qquad (15.2)$$

Where $\eta_i^e(\tilde{\omega}_{-i})$ is the expected rate of adoption within the neighbourhood of i. In the GNP model, the private idiosyncratic term H_i, is assumed invariable in time, but may differ from one agent to the other. It is useful to introduce the following notation for H_i - hereafter called *Idiosyncratic Willingness to Adopt* (IWA):

$$H_i = H + Y_i \text{ with : } \lim_{N \to \infty} \frac{1}{N} \sum_N Y_i = 0 \;\Rightarrow\; \lim_{N \to \infty} \frac{1}{N} \sum_N H_i = H \quad (15.3)$$

where Y_i is the outcome of an i.i.d. random variable Y with zero mean, distributed among the agents. Let $f_y(Y)$ be the Probability Density Function (pdf) of Y. As Y_i remains fixed, the resulting distribution of agents over the network of relations is a random field. Then, this model is formally equivalent to a "*Random Field Ising Model*" (RFIM) and the disorder is said to be "quenched" (i.e. there is no stochastic term). Therefore, agent's choices are purely deterministic (in contrast with the random utility approach in the DBB model, as mentioned before). An example of such model in sociophysics literature is (Galam 1997).

It is possible to relate our own model of binary choice with social influence to game theoretic models. Under our assumptions, all the agents have the same form of instrumental rationality (then, best response with respect to theirs expectations $\tilde{\omega}_{-i}$) and each agent has only two possible strategies: $\omega_i \in \Omega$. It is possible to represent the *total payoff* of an agent by the "normal form" matrix $G1$. From this standpoint, player 2 is a fictitious player; say a kind of *Neighbourhood Representative Player* (NRP), who stands for the behaviour of

---

[4] i.e. the price to buy one unit in the market case or some common cost in the non market case, cf. (Granovetter 1978, Glaeser and Scheinkman 2002)

the neighbourhood as a whole. If every k in the neighbourhood plays $\omega_k = 0$, the NRP plays the pure strategy $\omega_{nr} = 0$.

Table 15.1. Payoff matrix for an agent i and best reply equivalent potential game

(a) - game G1	$\omega_{nr} = 0$	$\omega_{nr} = 1$	(b) - game G2	$\omega_{nr} = 0$	$\omega_{nr} = 1$
$\omega_i = 0$	0	0	$\omega_i = 0$	$C - H_i$	0
$\omega_i = 1$	$H_i - C$	$H_i - C + J$	$\omega_i = 1$	0	$H_i - C + J$

Player i in rows - fictitious NRP - indexed nr - in columns

Conversely, if every k in the neighbourhood plays $\omega_k = 1$, the NRP plays the pure strategy $\omega_{nr} = 1$. However, the classical framework of two players game theory does not apply in numerous cases, because the strategic set of the player i and the NRP is generally asymmetric. Player i must plays only a pure strategy, while NRP can plays a mixed strategy. That is, the expected rate of adoption within the neighbourhood $\eta_i^e(\tilde{\omega}_{-i})$ corresponds to the expected share of $(\omega_k = 1)$ players in the neighbourhood. Consequently, player i plays his best response against the mixed strategy $\eta_i^e(\tilde{\omega}_{-i})$. Then, this later interpretation of the mixed strategy can be related to the framework of population games (Blume, 1997), where agent i plays in $N_{\vartheta i}$ bilateral confrontations against all agents k in their neighbourhood, with the payoff matrix G1' based on average payoff $((H_i - C)/N_{\vartheta i}$ for $\omega_i = 1$ against $\omega_k = 0$ and $(H_i - C + J)/N_{\vartheta i}$ for $(\omega_i = 1)$ against $(\omega_k = 1)$, respectively, zero otherwise). Indeed, since $N_{\vartheta i}$ is fixed, maximising the total surplus or the average surplus lead to the same solution.

One may add a constant term to one column and multiply all the columns by a constant term (here N_{ϑ_i}) without affecting the dominance ordering analysis, hence the best reply outcome. Thus, the following matrix in Table 15.1.b is said to be "best reply equivalent" to the one of Table 15.1.a This means notably that the Nash equilibria are the same whether one considers Game G1 (Table 15.1.a) or Game G1' (Table 15.1.b). This class of game with "best reply equivalence" (hence, similar Nash equilibrium) is called a (weighted) potential games (Monderer, Shapley, 1996).

For agents (type 0) such as: $H_i - C + J < 0$, strategy $\omega_i = 0$ (never adopt) is stricly dominant. Conversely, for agents (type 1) such as: $H_i - C > 0$, strategy $\omega_i = 1$ (always adopt) is strictly dominant. Then, the relevant situation is one with agents (type 3) such as $H_i - C < 0$ and $H_i - C + J > 0$. In this case, the choice depends on $\eta_i(\tilde{\omega}_{-i}^e)$, the expected rate of adoption within the neighbourhood. If all agents are of type (3), we have typically a coordination game with two Nash equilibrium; the so called "Stag Hunt Game". With bounded support for Y, $[Y_{min}, Y_{max}]$, this is the case if: $Y_{max} \leq C - H \leq Y_{min} + J$, what implies a sufficiently strength intensity of social effect, with respect to the dispersion of preferences $J \geq Y_{max} - Y_{min}$.

Table 15.2. A typology of interactions and demand dynamics

Neighbourhood	(a) No relations	(b) Localised relations	(c) Generalised relations
Level of interactions	(independent agents)	Localized interactions	Global interactions
sensitivity to the network topology	Null	Strong	Null
Avalanches	No	localised in the network	not localised in the network

15.2.2 Individual Interactions and Chain Reaction

In the first extreme case (a), there are no relations between agents. In this case, the aggregate demand depends on any interaction structure, and there is no external effect (local or global). The agents are independent one from each other. In the second extreme case (c), all agents interact by means of global interactions (e.g. the rate of adoption in the whole population). Let $\eta \equiv N_a/N$ be te rate of adoption within the population. For N sufficiently large, this rate is closed to the rate of adoption within the neighbourhood of each agents (full connectivity) say: $\eta \simeq N_a/(N-1)$. This case corresponds to the *means field approximation* in statistical physics. All agents are equivalent in the network. In this way, the aggregate demand is sensitive to the global external effect, but remains independent of the topology of the network (because the neighbourhood of each agent is composed of all the other agents). Thus, finite sequences of interdependent decisions called "avalanches" may arise, but such "dominoes effects" are *not localised* in the network of interactions and depend only on individual IWA, given the *global* rate of adoption, whatever the local rate of adoption (i.e. localized in the near neighbourhood). Finally, the intermediate case (b) corresponds to situations where agents have specified relations reified by the way of some network topology (regular neighbourhood or not). Interactions between agents are local, and the topology of the interpersonal network matters. This local interdependence may give rise to *localised avalanches* on the network (Table 15.2).

The term *avalanche* is associated with a chain reaction where the latter is directly induced by the behavioural modification of one or several other agents and not directly by the variation in cost. The cost influence is only indirect. For example in the left part of the (Table 15.3), an external cost variation (the same for all agents: C to C') induces a simultaneous (but independent of all social influence) change of two agents i and j (connected one to the other or not). Thus, the mechanism is directly related to the cost and independent of the social network. If, on the other hand, the cost variation induces the behavioural change of agent i, and therefore, because of agent i changes his behaviour, then agent j changes also his behaviour by social effect without any new change in cost, by "domino effect". In that case, the cumulative effect of a chain of such induced influences is called an "avalanche".

Table 15.3. Direct and indirect effect of prices upon individual choices

Direct effect of price	Indirect effect of price (social influence: avalanche)
Variation in cost $(C \longrightarrow C')$ ↙ ↘	Variation in cost $(C \longrightarrow C')$ ↓ Change of agent i ↓ Change of agent j
Change of Change of agent i agent j	

15.2.3 Avalanches and Hysteresis Loops in Aggregate Behaviour with Unique IWA

In this class of models, the adoption by a single "direct adopter" may lead to a significant change in the whole population through a chain reaction of "indirect adopters". The jump in the number of adopters occurs at different cost values according to whether the costs increases or decreases, leading to *hysteresis loops* as presented below. If the IWA is the same for all agents, ($H_i = H$, for all i), the model would be equivalent to the (quenched) Classic Ising Model with an "uniform external field": $H - C$.

In such a case, one would have a so called "first order transition", with all the population abruptly adopting as $H \geq C$. In Figure 15.1, this initial (decreasing) threshold is: $C_{min} = H$, where the whole population abruptly adopts. After adoption, the (increasing) cost threshold is: $C_{max} = H + J$, where the whole population abruptly choose $\omega_i = 0$ (for all i). When all agents are adopters, cost variations between: C_{min} and C_{max} have no effect on the agents choice. Within that zone $[C_{min}, C_{max}]$, there are two possible equilibria for a given cost.

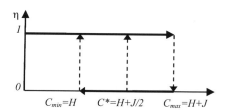

Fig. 15.1. Hysteresis with unique IWA ($H_i = H$)

From a theoretical point of view, there is a singular cost $C^* = H + J/2$ (the center of the interval $[C_{min}, C_{max}]$), which corresponds to the *unbiased*

210   Denis Phan and Stéphane Pajot

situation, where the willingness to adopt is neutral on average. Suppose that we start within such a neutral state. The agents makes their initial choice on the basis of some prior expectation about the number of adopters and further choice by updating this prior by use of the observed outcome. Assume first that all agents have the same expectation $\eta_i^e = \eta^e$ for all i. Then, each agent has a willingness to adopt equal to: $H + J\eta^e - C^* = J(\eta^e - 0.5)$. If $\eta^e > 1/2$, the expected surplus is positive and all agents adopt. Then, the ex post surplus will be $J/2$. Conversely, if: $\eta^e < 1/2$, the expected surplus is negative and no agent adopts. The final result is similar if we have two classes of people with heterogeneous expectations. Those with $\eta_i^{e+} > 1/2$ (in proportion α) adopt. If $\alpha > 1/2$, the percentage of adopters is such as pessimistic agents with $\eta_i^{e-} < 1/2$ but $\eta_i^{e-}\alpha > 1/2$ also adopt, and so on until complete adoption (and inverse process for $\alpha < 1/2$). This critical point plays a central role in the so called *spontaneous symmetry breaking*, even when agents are only locally connected. As in our simple example, the collective equilibrium state becomes identical to the individual state: either all agents adopt, or no agent adopts (Galam, 2004).

15.3 Avalanches and Hysteresis with Global and Local Interactions in Finite-size Population

This model describes the properties of many different systems (physical as well as social). It has been studied for various network architectures. In the presence of externality, and depending on the parameters, two different stable equilibria - or "phases" - may exist for a given cost: one with a small fraction of adopters (in some cases with no adopter) and one with a large fraction (in some cases, everybody adopts). By an external variation of the cost, a transition may be observed between these phases. Next subsection concerns the case of infinite size population and global interaction, while last subsection deals with both local and global interactions, by the way of computer simulations and finite size population.

15.3.1 Equilibrium Analysis: Phase Diagram with Global Externality

In order to present equilibrium results, let us consider now the special case of global externality from a static standpoint (e.g. without expectations). In this case, the individual surplus function (15.1) can be rewritten simplier as a function of the equilibrium value of the rate of adoption η.

$$W_i(\omega_i \,|\, \eta) \equiv \max_{\omega_i \in \{0,1\}} \{\omega_i(Y_i + H - C + J\eta)\} \qquad (15.4)$$

It is convenient to identify the *marginal adopter*, indifferent between adopting and not adopting. Let $H_m = H + Y_m$ be his idiosyncratic willingness to adopt (IWA). This marginal adopter has zero surplus $W_m = V_m = 0$, $\forall \omega_m \in \Omega$, that is:

$$Y_m = C - H - J\eta \qquad (15.5)$$

Consequently, an agent adopts if $Y_i > Y_m$ and does not adopt otherwise. Then, if the law of Y has a continuous pdf, the rate of adoption is the solution of the following:

$$\eta = P(Y_i > Y_m) \equiv G_y(Y_m) \equiv \int_{Y_m}^{\infty} f_y(y)\, dy \qquad (15.6)$$

More specifically, assume that Y follows a symmetric triangular law, with bounded support $[-a, +a]$. The fixed point condition (15.6) has one or three solutions, with two stable equilibria in this later case (Phan, Semeshenko 2006). More generally, this is a generic property of this model of binary choice with externality for a large class of distribution (Gordon *et al.* 2006).

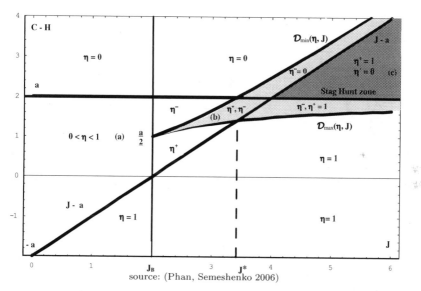

Fig. 15.2. Equilibrium regimes in the phase spaces: $(J, C - H, a = 2)$

According to a methodology introduced by the Physicists, Figure 15.2 exhibits in the phase plane $(J, C - H)$ a cartography of regions with one equilibrium or two equilibria, with respect to the value of corresponding parameters. For the detail of the calculus for this triangular case, see (Phan, Semeshenko, 2006). A stable equilibrium can be viewed as a Nash equilibrium of a population game (section 15.2.1). For low cost and sufficiently strength of social coupling, everybody adopt (in zone south and south east on the phase diagram).

Conversely, for high cost and weak social coupling, nobody adopt (North West). In the west and south west, for $J < J_B$ and $J - a < C - H < a$, there is a *polymorphic Nash equilibrium* with both non-adopters and adopters in proportion $0 < \eta < 1$ (Figure 15.3.a). Let $\mathcal{D}(\eta, j) \equiv G^{-1}(\eta) - J\eta$. For $J > J_B$, there is a zone with two stable Nash equilibria (the grey zone on Figure 15.2). This zone is delimited by two frontiers given by $\mathcal{D}_{min}(\eta, j) = J - a + a^2/(2J)$ and $\mathcal{D}_{max}(\eta, j) = a - a^2/(2J)$. Therefore, if: $\mathcal{D}_{min}(\eta, j) > C - H > \mathcal{D}_{max}(\eta, j)$ there are two equilibria: one rate of adoption less than 50% (possibly 0%) and other more than 50% (possibly 100%) .

Accordingly, the polymorphic single equilibrium zone has two extensions for $J^* > J > J_B$, with: $0 < \eta^- < 0.5$ if: $\mathcal{D}_{min}(\eta, j) < C - H < a$ and: $0 < \eta^+ < 0.5$ if: $J - a < C - H < \mathcal{D}_{max}(\eta, j)$ respectively. In the darker grey zone, the strength of social coupling is such as $J > 2a$ and therefore: $a \leq C - H \leq J - a$. According to section 15.2.1 all agents are of type (3), and we have a "Stag Hunt" coordination game with two equilibria, one without any adoption and another with complete adoption.

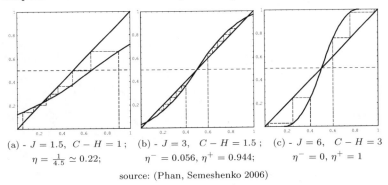

(a) - $J = 1.5$, $C - H = 1$;   (b) - $J = 3$, $C - H = 1.5$;   (c) - $J = 6$, $C - H = 3$
$\eta = \frac{1}{4.5} \simeq 0.22$; $\eta^- = 0.056$, $\eta^+ = 0.944$; $\eta^- = 0$, $\eta^+ = 1$

source: (Phan, Semeshenko 2006)

Fig. 15.3. Sequential dynamics with a triangular distribution of IWA

In order to illustrate some typical equilibrium cases from the phase diagram, let us consider a recurrent relation drawn from the fixed point condition (15.6) in the case of the agents have identical myopic expectations: $\eta^e(t) = \eta(t - 1)$. Then, $Y_m(t) = C - H - J\eta(t - 1)$. This recurrent relation allows us to represent agents' learning by a graphic of fixed point dynamics on Figure 15.3. In Figure 15.3.a the stable equilibrium is unique, while there are two stable equilibria separate by an instable fixed point on Figure 15.3.b (polymorphic) and Figure 15.3.c (Stag Hunt).

15.3.2 Avalanches and Hysteresis Loops in Aggregate Behaviour with Logistic IWA

The previous results concern the case of infinite size population and global interaction. This section is devoted to the case of a finite size population by the way of computer-based simulations. From work in progress, we present some sample experiments with both global and local interactions on a *random field* with logistic quenched disorder, representing idiosyncratic fixed IWA.

In the presence of a quenched disorder, the number of customers may evolve by a serie of cluster flips, or avalanches. If the disorder is strong enough (i.e. the variance σ^2 of Y is large with respect to the strength of the social coupling J), there will be only small avalanches (There are numerous agents following their own H_i). If σ^2 is small enough, the phase transition occurs through a unique "infinite" avalanche, similar to the case with the unique H for all agents (section 15.2.3). This is called a "first order phase transition" by physicists. In intermediate regimes, a distribution of smaller avalanches of various sizes can be observed. It is useful to consider as exemple a sample of a simulation, using the multi-agent framework Moduleco-Madkit (Gutknecht and Ferber 2000; Phan 2003; Michel *et al.* 2005) [5].

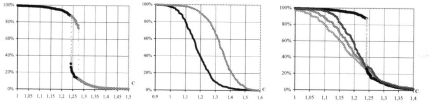

a - Global externality (TC)   b - Circle, Neighbours = 2   c - upstream branch: circle and TC
$H = 1$, $J = 0.5$, $\beta = 10 \Rightarrow \sigma = \pi.\sqrt{3}/10 \simeq 0.544$, Logit pseudo-random generator, seed = 190
(a-b) upstream (black)and downstream (grey) trajectories (c) circle with $N = 2, 4, 8$ and TC:
from grey to black

Fig. 15.4. Hysteresis in the trade-off between cost and adopters under synchronous activation regime (Moduleco-Madkit: 1296 agents - synchronous activation regime)

Figure(s) 15.4.a-c shows for a set of particular experience with the same distribution of IWA (seed = 190). Points are equilibrium rate of adoption for the whole system for cost incremented in steps of 10^{-4} under the synchronous activation regime (all agents update their behaviour at the same time). One observes a hysteresis phenomenon with phase transitions around the theoretical point of symmetry breaking: $C^* = H + J/2 = 1.25$. Figures 15.4.a deals with the "global" externality, while Figure 15.4.b corresponds to a "local" externality (on a one-dimensional periodic lattice: the circle case, with two nearest neighbours) with the same parameters and IWA distribution in both cases. Figure 15.4.d shows the upstream branch (decreasing costs) of a circle with nearest neighbours ($N = 2, 4, 8$) and the same global externality case (TC) than on Figure(s) 15.4.a. Figure 15.4.a shows the details of straight hysteresis corresponding to the "global" externality (complete connectivity). In this case, the trajectory is no longer gradual, like in the local interdependence case on Figure 15.4.b. Along the upstream equilibrium trajectory (with decreasing costs) an avalanche arises for $C = 1.2408$, by a succession of cluster

---

[5] For the simulations presented below, we have a logistic distribution where $\beta = \pi.\sqrt{3}/\sigma$ is the logistic parameter, $H = 1, J = 1/2$ and $\beta = 10$; ($J.\beta = 5$)

flips, driving the system from an adoption rate of 30% towards an adoption rate of roughly 87%. Along the downstream trajectory (with increasing costs) the externality effect induces a strong resistance of the system against a decrease in the number of adopters. The phase transition threshold is here around $C = 1.2744$. At this threshold, the equilibrium adoption rate decreases dramatically from 73 % to 12.7 %.

The *threshold of exposure* (TE) is the proportion of adopters in the local neighbourhood of an agent sufficient enough to induce a change in his behaviour (Valente 1995). For finite neighbourhood, this TE evolves by discrete jump and therefore it is very sensitive to the size of the neighbourhood. This threshold effect may be either favourable or unfavourable to adoption, depending of the relative position of the agent with respect to the *unbiased situation*. For instance let $J = 1$ and $C = 1$. The unbiased situation is such as: $C^* - H = J/2$, hence $H = 0.5$. If $H_i = 0, 4$ (the agent i is below the unbiased situation), then $C - H_i = 0.6$; for $N = 2$ the TE_2 is 2, say 100%; while $TE_4 = 3$ (75%) with $N = 4$ and $TE_8 = 5$ (62,5%) with $N = 8$. Thus, in this case, the relative TE (i.e. the rate of the TE over the neighbourhood) decreases with the widening of the neighbourhood. Conversely, if the IWA is such as the agent is above the unbiased situation say, $H_i = 0, 6$: then $C - H_i = 0.4$, the TE_2 is 1 (50%) for $N = 2$. This rate remains the same with $N = 4$ ($TE_4 = 2$) and with $N = 8$ ($TE_8 = 4$). In this later case, we need a neighbourhood equal or superior to $N = 10$ in order to reduce the relative TE below the relative threshold of 50%. The finite size effect of the TE is then both discontinuous and asymmetric.

For finite size population and finite neighbourhood, the equilibria distribution is very sensitive to the possibility of local clusters both with higher or lower adoption with respect to the mean field case (complete connectivity or social influence). This is related to both the discrete distribution of the thresholds and the possibility of extreme situation (where an agent is surrounded by neighbours all with either a small or a great IWA). Such effect is more sensitive for low cost / high degree of adoption, where the adoption is slower with local neighbourhood, due to the existence of clusters of non-adopters, called "frozen zone".

Figure 15.4.c shows the evolution of the rate of adoption for several configurations of the network: one dimensional periodic (circle) with near-neighbourhood of size 2, 4, 8 and complete connectivity (from light grey to black respectively). In this case with 1296 agents, the negative effect of local interdependence is clearer than the positive one (for low rates). In the case under consideration, the widening of the neighbourhood has a little positive effect on adoption. For relatively high cost / low rates of adoption, the number of adopters is higher than for the full connectivity. For relatively small costs (high rates of adoption), the existence of local interdependences (frozen zone) has a strong negative effect, hence the number of customers is clearly smaller than in the case of global influence, but this later effect is little attenuated by the widening of the neighbourhood effect.

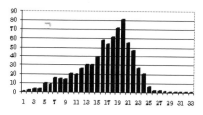 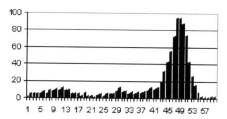

a - the avalanche for p = 1.2408 (seed = 190)   b - avalanche for p = 1.2415 (seed = 40)
parameters: $H = 1$, $J = 0.5$, $\beta = 10$ Moduleco: synchronous activation regime).

Fig. 15.5. Examples of chronology and sizes of induced adoptions in the avalanche at the phase transition under global externality in two single experiences with 1296 agents

In the case of a finite size sample, there is some local irregularities in the discrete distribution of characteristics (IWP), even with "near-perfect" pseudo-random generator. Therefore, the shape of an avalanche is completely dependent on the realizations of Y_i. Then, gaps in the ordered sequence of the Y_i produce fluctuations in the chronology of induced adoptions, as well as possible multi-modal shape, like in Figure 15.5.b. Despite the non-generic properties of such figures, this kind of historic profile remains relevant for empirical experiments in finite size situations.

Figure 15.5.a shows the chronology of an avalanche in the case of the upstream branch of the equilibrium trajectory, for $C = 1.2407$. The evolution follows a smooth path, with a first period of 19 steps, where the initial change of one customer leads to growing induced effects from size 2 to size 81 (6.25 % of the whole population). After this maximum, induced changes decrease in 13 steps, including 5 of size one only. Figure 15.5.b shows a different case, with more important induced effects, both in size and in duration (seed 40). The initial impulsion is from a single change for $C = 1.2415$ with a rate of adoption of 19.75 %. The first wave includes the first 22 steps, where induced changes increase up to a maximum of 11 and decrease towards a single change. During this first sub-period, 124 agents change (9.6 % of the whole population). After step 22, a new wave arises with a growing size in change towards a maximum of 94 agents both in step 48 and 49. The total avalanche duration is 60 steps, where 924 induced agent changes arise (71 % of the population - 800 in the second wave). As suggested previously, the steepness of the phase transition increases when the variance σ^2 of the logistic distribution decreases (increasing β).

The closer the preference of the agents to each other, the greater the size of avalanches at the phase transition (Figures 15.6.a-b). Figure 15.6.c shows a set of upstream trajectories for different values of β taken between 20 and 5 ($10 \geq J\beta \geq 5$), in the case of global externality. The scope of the hysteresis decreases with β ; for $\beta < 5$, there is no longer any hysteresis at all (remark that intermediate positions in straight hysteresis are transitory equlibibrium

216   Denis Phan and Stéphane Pajot

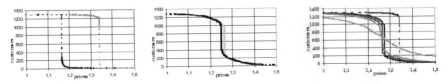

a - large hysteresis for $\beta = 20$ b - narrow hysteresis for $\beta = 9$ c - upstream branch with $20 \geq \beta \geq 5$
(a - c): total connectivity (TC)

Fig. 15.6. The trade-off between cost and adopters (synchronous activation regime)

(in light grey in   15.4.a) and finite size effect, and do not appear in the analytical case with "infinite" population.

Acknowledgements

The present paper is based on part of a previous unpublished work with J.P. Nadal, presented at CEF 2003 in Seattle (USA). The authors acknowledge M. B. Gordon, and J.P. Nadal and V. Semeshenko for their significant contribution to the development of the GPN model and Roger Waldeck for Valuable Remarks, as well as the joint program "Complex Systems in Human and Social Sciences" of the French Ministry of Research and of the CNRS for financial support of the project 'ELICCIR'. D.P. is CNRS member.

References

[1] Becker G.R. (1974) A Theory of Social Interactions, Journal of Political Economy, 82(6) pp. 1063-1093
[2] Blume L.E. (1997) Population games in Arthur et al. (eds.) op cit. pp. 425-460
[3] Blume L.E., Durlauf S.N., (2001) The Interaction-Based Approach to Socioeconomic Behavior, in Durlauf, Young, (eds.) op.cit., pp. 15-44
[4] Block H., Marschak J. (1960) Random Orderings and Stochastic Theories of Response, in: Olkin I. ed. Contributions to Probability and Statistics, I, Stanford University Press
[5] Brock W.A., Durlauf S.N. (2001a) Discrete choice with social interactions, Review of Economic Studies, 2001,68, pp. 235-260
[6] Brock W.A., Durlauf S.N. (2001b) Interaction Based Models, In: Heckman, Leamer (eds.) Handbook of Econometrics Vol. 5, Ch54, Elsevier Science, pp. 3297-3380
[7] Bourgine P., Nadal J.P. eds., (2004) Cognitive Economic, Springer Verlag, Berlin, N.Y.

[8] Durlauf S.N., (1997) Statistical Mechanics Approaches to Socioeconomic Behavior, In: Arthur, Durlauf, Lane, (eds.) The Economy as an Evolving Complex System II, Santa Fe Institute Studies in the Sciences of Complexity, Vol.XXVII, Addison-Wesley, pp. 81-104

[9] Durlauf S.N., (1999) How can Statistical Mechanics Contribute to Social Science ?, Proceedings of the National Academy of Sciences 96:10582-10584

[10] Durlauf S.N., Young P., (2001) (eds) Social dynamics, MIT Press, Cambridge Ma.

[11] Föllmer H., (1974) Random Economies with many Interacting Agents, Journal of Mathematical Economics, 1:51-62

[12] Galam S., Aharony A. (1980) A New Multicritical Point in Anisotropic Magnets. I. Ferromagnets in a random longitudinal Field, Journal of Physics C, 15, pp. 529-545

[13] Galam S., Aharony A. (1981) A New Multicritical Point in Anisotropic Magnets. II. Ferromagnets in in a random skew field, Journal of Physics C, 14, pp. 3603-3619

[14] Galam S. (1982) A New Multicritical Point in Anisotropic Magnets. III. Ferromagnets in both a Random and a Uniform Longitudinal Field, Journal of Physics C, 15 pp. 529-545

[15] Galam S., (2004) Spontaneous Symmetry Breaking, in Bourgine, Nadal. (eds.) op. cit. pp. 157-168

[16] Galam S., Gefen Y., Shapir Y., (1982) "Sociophysics: A Mean Behavior Model for the Process of Strike", Mathematical Journal of Sociology, 9:1-13

[17] Galam S. (1997) Rational group decision making: A random Field Ising model at T=0, Physica A, 238, pp. 66-80

[18] Glaeser E.L., Sacerdote B., Scheinkman J.A. (1996) Crime and social interactions, Quarterly Journal of Economics, 1996,Vol. CXI, pp. 507-548

[19] Glaeser E.L. Scheinkman J.A. (2002) Non-market interactions, in: Dewatripont M. Hansen L.P. Turnovsky S. Advances in Economics and Econometrics: Theory and Applications, Eight World Congress, Cambridge University Press

[20] Gordon M.B., Nadal J.P., Phan D., Vannimenus J. (2005) "Seller's dilemma due to social interactions between customers", Physica A , N356, Issues (2-4):628-640

[21] Gordon M.B., Nadal J.P., Phan D., Semeshenko V. (2006) Discrete Choices under Social Influence: Generic Properties. In : WEHIA 2006, 1st International Conference on Economic Sciences with Heterogeneous Interacting Agents, 15-17 June, Bologna

[22] Granovetter M., (1978) "Threshold Models of Collective Behavior", American Journal of Sociology, 83(6):1360-1380

[23] Gutknecht, O. Ferber, J. (2000) Madkit: a Generic Multi-Agent Platform,Autonomous Agents (AGENTS 2000), Barcelona, ACM Press,pp. 78-79

[24] Ioannides Y.M. (2006) Topologies of Social interactions, Economic Theory, 28 p.559-84

[25] Kindermann R., Snell J.L., (1980) On the Relation between Markov Random Fields and Social Networks, Journal of Mathematical Sociology, 7:1-13

[26] Kirman A.P. (1983) Communications in Markets: A suggested Approach, Economic Letters 12 pp. 1-5.

[27] Kirman A.P. Oddou C., Weber S. (1986) Stochastic communication and coalition formation, Econometrica, 54 pp. 129-138.

[28] Marschak J. (1960) Binary Choice Constraints on Random Utility Indicators, in K. Arrow ed. Stanford Symposium on Mathematical Methods in the Social Sciences, Stanford University Press

[29] Mansky, C. (1977) The Structure of Random Utility Models, Theory and Decision 8, pp. 229-254.

[30] McFadden D. (1974) Conditional Logit Analysis of Qualitative Choice Behavior, in Zarembka ed. Frontiers in Econometrics, Academic Press, N.Y.

[31] McFadden D. (1976) Quantal Choice Analysis: A Survey, Annals of Economic and Social Measurement 5(4), pp. 363-390

[32] Michel F., Daniel G., Phan D. & Ferber J. (2005) Integration Moduleco-Madkit: first results. In: SMAGET 05, Joint Conference on Multi-Agent Modelling for Environmental Management, March, 21-25

[33] Monderer D., Shapley L. (1996) Potential Games, Games Econ. Behaviour 14:124-143.

[34] Nadal J.P., Phan D., Gordon M.B., Vannimenus J. (2005) Multiple equilibria in a monopoly market with heterogeneous agents and externalities, Quantitative Finance, December Vol 5(6):1-12

[35] Phan D. (2004) From Agent-Based Computational Economics towards Cognitive Economics, In: Bourgine, Nadal. (eds.) op. cit. pp. 371-398

[36] Phan D., Gordon M.B & Nadal J.P. (2004) Social Interactions in Economic Theory: an Insight from Statistical Mechanics, In: Bourgine, Nadal. (eds.) op cit., pp. 225-358

[37] Phan, Semeshenko, (2006) Multiplicity of Equilibria in Binary Choice Model with Social Influence: the Triangular distribution case, in revision for EJESS

[38] Schelling T.C., (1973) Hockey Helmets, Concealed Weapons and Daylight Saving A study of binary choices with externalities, Journal of Conflicts Resolution, September 17(3):382-428

[39] Schelling T.C., (1978) Micromotives and Macrobehavior, Norton and Compagny, N.Y.

[40] Sethna J.P., Dahmen K., Kartha S., Krumhansl J.A., Roberts B.W., Shore J.D. (1993) Hysteresis and Hierarchies: Dynamics of Disorder-

Driven First-Order Phase Transformations, Physical Review Letters, 70:3347-3350

[41] Sethna J.P., Dahmen K.A., Perkovic O. (2006) Random-Field Ising Models of Hysteresis in The Science of Hysteresis Vol. II, G. Bertotti, I. Mayergoyz (eds), Elsevier, Amsterdam

[42] Thurstone L.L. (1927) A Law of Comparative Judgement, Psychological Review, 34, pp.273-286.

[43] Stanley H.E. (1971) *Introduction to phase transitions and critical phenomena*, Oxford University Press.

[44] Valente T., (1995) Networks Models of the Diffusion of Innovations, Hampton Press Inc

[45] Weidlich W., Haag G., (1983) Concepts and Models of a Quantitative Sociology, the Dynamics of Interacting Populations, Springer Verlag, Berlin, New York

16

Dynamics of a Public Investment Game: from Nearest-Neighbor Lattices to Small-World Networks

Roberto da Silva[1], Alexandre T. Baraviera[2], Silvio R. Dahmen[3], and Ana L. C. Bazzan[1]

[1] Instituto de Informatica, UFRGS, {rdasilva,bazzan}@inf.ufrgs.br
[2] Instituto de Matematica, UFRGS, baravi@mat.ufrgs.br
[3] Instituto de Fisica, UFRGS, dahmen@if.ufrgs.br

Summary. In this work we analyze the time evolution of the wealth of a group of agents in a public-investment-game scenario. These are part of a small-world network, where connections depend on a probability p and investment depends on a binary variable σ (motivation). This variable tries to emulate one's perception of other players' actions. We study the effect of the connectivity on the wealth of the group as well as the dynamics when idyosincratic types are introduced in the game.

16.1 Introduction

The list of publications on nontrivial phenomena which arise due to the interplay between microscopic (individual) rules and macroscopic (group) behavior in the fields of complex and multiagent systems is extensive. In the context of socioeconomic behavior, this has been thoroughly discussed by Durlauf [4]. Within this scenario we present a variation of a simple "public investment game" [2]. In its original version, one wishes to model public spending on public goods. Players can invest their money in a common pool, and profits are equally distributed among all participants irrespective of their contributions. Clearly it would be "fair" for people with similar amounts of money to invest similar quantities. However individuals are different: each player, blind as to what regards others' contributions, would default and invest nothing if it were rational. For purely rational players the dominant solution is to default.

To give the model a more realistic flavor we let agents interact and invest according to the actions of their immediate neighbors. We do this by introducing a binary variable we call *motivation*, and whose update depends on a random variable [7]. The aim is to simulate natural causes which might affect the way players assess the investment. The return per agent is considered to be a function of the average investment, in close relation to cooperative game-theory.

We start with periodic boundary conditions (players in a ring). To explore more complex networks which allow agents to be influenced by others far away from them, we also consider small-world networks [8]. Such networks can be built from regular ones by giving each player a probability p of reconnecting to another player chosen randomly.

In this paper we explore the dynamics of a fraction of agents who operate at a deficit, given that each player starts the game with the same quantity of money. We also measure the probability of a particular agent not losing all its money up to time t as function of parameter p of the small-world. We study the non-trivial dynamics that emerges out of this system by looking for the density of motivated agents in the model and determining a phase diagram.

The paper is organized as follows: In section 16.2 we discuss other game-theoretic settings which use the small-world metaphor. In section 16.3 we define our model. In section 16.4 we study a case for which an exact solution exists, namely in a complete graph. In section 16.5 we extend our studies to small-world networks with arbitrary coordination number K. We also study how the introduction of idiosyncratic agents into the system affects the dynamics of the global wealth. We close the paper with section 16.6.

16.2 Groups as Networks of Agents in Small-World Scenarios

It is easy to recognize that networks of coupled individual elements are not only a paradigm for studying artificial systems, but also an artifact that appears often in Nature and social systems. Watts and Strogatz [8] studied networks of coupled elements through an analogy with the *small-world phenomenon* (SW), which is based on the fact that in large societies there is normally a shortcut between any two persons. A classical parallel is the "six degrees of separation" concept [5]: there is a path of acquaintances with typical length of six between most pairs of people in the United States.

In a work linking the SW and the Iterated Prisoners' Dilemma (IPD) of Nowak and May [6], Abramson and Kuperman [1] started out with a ring of agents (*i.e.* those who only see their immediate left and right neighbors) and allowed them to interact with agents located somewhere in the network. This is represented by a graph with N vertices, each one connected to K vertices. After an interaction, a "rewiring" process takes place for each vertex v_i with probability p and some connections are broken and replaced by connections to vertices far from v_i. p is thus a measure of the regularity of the network.

In the setting proposed by Abramson and Kuperman, N players are connected, on average, to $2K$ vertices (after rewiring). A round of the game consists of the interaction of every player with all neighbors and the sum of the points collected based on a given payoff matrix. Strategies are to cooperate or defect. After the interaction with all players in the "neighborhood" (here not necessarily only the closest neighbors), players are allowed to inspect the profits collected by neighbors and imitate the one which has collected more points as in [6], where the final rate of cooperation (simply the percentage of agents whose tactic is cooperate divided by the total number of agents) is 32%. When the SW links are introduced, the rate of cooperation is influenced by p.

In [3], the focus is on analyzing the performance of a society composed of agents playing the IPD in the presence of agents with attachment to others. Altruistic agents are interested in the good performance of their group as a whole, as well as on their own, since the social group provides also a base for support in case the agent itself is not performing well. Moreover the best performance in the group will be imitated. Besides altruistic agents, the society can also be populated with egoistic ones. In the context of the public investiment game this was done in [7], with the result that the dynamics can be completely different whether one has a larger percentage of altruistic or selfish agents.

16.3 The Model in a Small-World Network

We consider a game of L investors or economic agents which, starting the game with a quantity w_0 of money, can invest a quantity S_i. Agents invest cooperatively, *i.e.* the average profit of the group influences the investment motivation level of each agent. This is modelled by a binary variable σ_i where $\sigma_i = 1$ means an agent is motivated while $\sigma_i = 0$ means it is not. This abstraction aims at capturing issues such as insider information and economic prospects as perceived by agents.

An agent $i = 1, ..., L$, in a small-world network built from a regular lattice with arbitrary coordination K has a set of neighbors we denote ξ_i. We define the invested quantity S_i through

$$S_i(t) = \sigma_i(t) + F(\rho_i(t)) \tag{16.1}$$

where the function $F(\rho_i(t))$ depends on the density of motivated neighbors of agent i at instant t, $\rho_i(t) = (1/|\xi_i|) \sum_{j \in \xi_i} \sigma_j(t)$, as follows:

$$F(\rho_i(t)) = \begin{cases} v_1(t) & \text{if } \rho_i(t) > 1/2 \\ v_2(t) & \text{if } \rho_i(t) = 1/2 \\ v_3(t) & \text{if } \rho_i(t) < 1/2 \end{cases}$$

where $v_1(t)$, $v_2(t)$ and $v_3(t)$ are arbitrary functions and $|\xi_i|$ is the cardinality of set ξ_i. For the sake of clarity we restrict our investment to four possibilities $S_i \in \{0, 1, 2, 3\}$, *i.e.*

$$v_l(t) = 3 - l \tag{16.2}$$

Table 16.1. Investment rules relating motivation levels to investment

σ_i (motivation)	ρ_i (density of neighborhood)	S_i (investment)
0	$< 1/2$	0
0	$= 1/2$	1
0	$> 1/2$	2
1	$< 1/2$	1
1	$= 1/2$	2
1	$> 1/2$	3

where $l = 1, 2, 3$ (see table 16.1).

To update the motivation level of the agents, we represent the average investment of agents in the t−th iteration as:

$$S(t) = \frac{1}{L} \sum_{k=1}^{L} S_k(t) \tag{16.3}$$

where periodic boundary conditions are imposed and S_k is given by (16.1).

We assume that the overall profit is modulated by a random variable r (noise) uniformly distributed in $r \in [-1, 1]$. The return per agent is given by:

$$g_k(t) = (a + br) S(t) - S_k(t) . \tag{16.4}$$

When $b = 0$ we have the deterministic (or noiseless) case. On the other hand, if $a = 1$ and $b = 1/2$, profits $(0 < r < 1)$ and losses $(-1 < r < 0)$ are allowed only within a range which depends on the average investment $S(t)$. Individually agents can be better off or not. At any given time t an agent has an accumulated wealth given by

$$W_k(t + 1) = W_k(t) + g_k(t) \tag{16.5}$$

where $W_k(1) = w_0$, $k = 1, ..., L$. We update the motivation at each time step by the profit rate $g_k(t)$:

$$\sigma_k(t + 1) = \begin{cases} \frac{1}{2}\left(1 + \frac{g_k(t)}{|g_k(t)|}\right) & \text{if } g_k(t) \neq 0 \\ \\ 0 & \text{otherwise} \end{cases} \tag{16.6}$$

This update is based on a simple principle: an agent's wealth relies on the wealth of the group. However, since agents are autonomous and there is room for cheating, we end up with two kinds of situations: one in which everyone is cooperative, and another where different types of individual behavior can be simulated. Before we explore the dynamics of our model we discuss in the next section a case where analytical results can be obtained.

16.4 The Game in a Complete Graph

We consider the noiseless scenario ($b = 0$) and all agents connected to one another. If we call

$$\rho = \frac{1}{L} \sum_{k=1}^{L} \sigma_k$$

the mean motivation then for a large number of agents ($L \to \infty$) the difference from ρ to any one of the local motivations ρ_i is less than $1/L$. In this case the approximation $\rho_i = \rho$ can be used.

With this in mind and with the help of the real function h such that $h(x) = 0$ if $x < 0$, $h(x) = 2$ if $x > 0$ and $h(0) = 1$ we can rewrite $S_i(t)$ as

$$S_i(t) = \sigma_i(t) + h(\rho_i(t) - 1/2) = \sigma_i(t) + h(\rho(t) - 1/2)$$

where in the last inequality we used the large L assumption. The return per agent is now given by:

$$g_k(t) = aS(t) - S_k(t) = a\rho(t) + (a - 1)h(\rho(t) - 1/2) - \sigma_k(t)$$

Fixed points

The first question we want to address is: how does $\rho(t)$ behave? We first recall that $\rho = 0$ is a fixed point of the system's dynamics irrespective of the value of a. For $\rho = 1$ to be a fixed point we need $\sigma_k = 1$ for all k at all times, $i.e.$ we need always a positive return g_k. One is left with the task of determining the correspoding values of a for this to happen. Before we proceed to these cases, we note that another interesting point is $\rho = 1/2$ for $a \in (2/3, 4/3)$. In this case the return per agent is:

$$g_k = \frac{3}{2}a - 1 - \sigma_k$$

If $\sigma_k = 0$, then $g_k > 0$ and $0 \to 1$. For $\sigma_k = 1$ then $g_k < 0$ and $1 \to 0$. Hence, the two populations (motivated and unmotivated) exchange their role, but the *density* ρ is constant, despite the oscillatory behavior of each individual agent.

The $\rho = 0$ Attractor

We now determine the cases for which $\rho_0 = 0$ is an attractor. For this to be the case we require $g_k \leq 0$, since then some σ_k's will flip to 0. But $g_k \leq 0$ is

$$a\rho + (a - 1)h(\rho - 1/2) - \sigma_k \leq 0$$

To verify the condition above for any value of σ_k it is sufficient that

$$a\rho + (a - 1)h(\rho - 1/2) \leq 0$$

There are three cases to consider: If $\rho < 1/2$ then $h(\rho - 1/2) = 0$ and so the condition is $a\rho \leq 0$. If $\rho = 1/2$ then $h = 1$ and this way we have $a\rho + (a-1) \leq 0$ (i.e. $\rho \leq \frac{1}{a} - 1$). Since $\rho = 1/2$ this gives us $a \leq 2/3$. The third case is $\rho > 1/2$. Now $h = 2$ and then the condition is $a\rho + (a - 1)2 \leq 0$ (i.e. $\rho \leq \frac{2}{a} - 2$).

The information above gives an immediate basin for the $\rho = 0$ attractor, but the basin itself can be larger. For example, fixing $a \in [0, 2/3]$ and taking $0 < \rho < 1/2$, the return per agent is:

$$g_k = a\rho + (a - 1)h(\rho - 1/2) - \sigma_k = a\rho - \sigma_k$$

Those agents for which $\sigma = 0$ the return is positive: they change their motivation to 1; those for which $\sigma = 1$ the return is negative and the motivation becomes 0. Hence, the density changes to $1 - \rho > 1/2$ and in one more interaction the point $\rho = 0$ is reached.

The same reasoning can be applied to the case $a \in (2/3, 4/5]$, with $0 < \rho < 1/2$. The conclusion is that for all situations where $\rho \geq 3 - 2/a$ we have the dynamics $0 \rightarrow 1$ and $1 \rightarrow 0$ as above, showing that the density goes to $1 - \rho$. This again corresponds to the immediate basin of $\rho = 0$.

The $\rho = 1$ Attractor

Now we look for situations in which the attractor is the configuration where all the agents remain motivated forever, i.e., $\sigma_k(t) = 1$ for all k and all t, corresponding to $\rho = 1$.

In order to change from $\sigma_k = 0$ to $\sigma_k = 1$ we need $g_k > 0$, in other words:

$$a\rho + (a - 1)h(\rho - 1/2) - \sigma_k > 0$$

Again we would like to have the inequality above satisfied independently of σ_k. For this it is sufficient to that:

$$a\rho + (a - 1)h(\rho - 1/2) - 1 > 0$$

If $\rho < 1/2$ this is equivalent to $\rho > 1/a$; if $\rho = 1/2$ this gives $a > 4/3$. For $\rho > 1/2$ then the condition is $\rho > \frac{3}{a} - 2$.

As in the previous section, this gives the immediate basin of $\rho = 1$. Now take $\rho < 1/2$ and such that $\rho < 3 - 3/a$ (for $a \geq 1$); if $\sigma_k(t) = 0$ then $g_k = a\rho > 0$ showing that $0 \rightarrow 1$ and so the new density is at least $1 - \rho$, being on the basin of $\rho = 1$.

Oscillatory Cases and the Phase Diagram

We now try to find distinct kinds of behavior for the evolution of ρ in the remaining part of the $a \times \rho$ diagram.

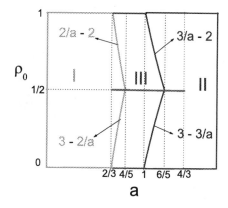

Fig. 16.1. Phase diagram of the public investment game in the complete graph, where each agent interacs with all $(L-1)$ remaining agents. Regions I, II and III correspond respectively to fixed points $\rho = 0$, $\rho = 1$ and oscillatory behavior. The fixed point $\rho = 1/2$ happens for $a \in [2/3, 4/3]$.

If we start with $\rho < 1/2$, then the dynamics is $0 \to 1$ and $1 \to 0$, showing that the density goes to $1 - \rho$. Now, if $\rho > 1/2$ the dynamics in the region is again $0 \to 1, 1 \to 0$. Hence, each agent oscillates with period 2 and the density is oscillatory with period 2, going from ρ to $1 - \rho$ and back to ρ. Finally we show a phase diagram where all situations are explored. In region **I**, for all initial (a, ρ_0) the fixed point $\rho = 1$ is reached, while the same happens for $\rho = 0$ in region **III**. Region **II** represents the oscillatory behavior. The straight line $\rho_0 = 1/2$ for $a \in [2/3, 4/3]$ represents the fixed point $\rho = 1/2$, which is a special case of the more general oscillatory phase: half of the motivated (unmotivated) agents change their motivation, becoming unmotivated (motivated).

16.5 Small-World Networks - Numerical Analysis

In this section we analyze the more general case of small-world networks. As these case do not allow for an analytic solution, we present the results of our simulations in what follows.

16.5.1 Numerical Phase Diagram for the Deterministic Case (b = 0) in the SW Network

We analyzed the time evolution of $\rho(t)$ for small-world networks when $b = 0$. In this case we considered values of $a = 0, ..., 2$ and $\rho_0 = 0, ..., 1$, using the steps of size $\Delta a = 0.02$ and $\Delta \rho_0 = 0.01$, **i.e.** $\rho_0^{(k)} = k\Delta\rho_0$ and $a^{(k)} = k\Delta a$,

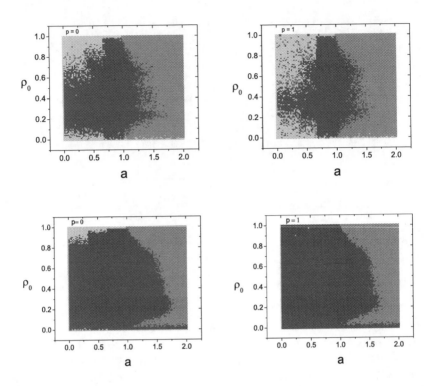

Fig. 16.2. Phase diagram of the public investment game without noise (b=0) in a small world network for extremes cases p = 0 and p = 1 (Upper: 1 run; lower: 60 runs). The dark gray region depicts the oscillatory behavior, while light gray and gray mean the regions of fixed points $\rho = 0$ and $\rho = 1$ respectively.

where $k = 0, ..., 100$. In this situation we observed the extreme cases in the small worlds $p = 0$ and $p = 1$ the intermediate cases only showing a transition between these two cases. We also discuss two situations: a) the phase diagram with only one run and b) over a sample of 60 runs.

We have implemented an algorithm that identifies, for each pair $(a^{(k)}, \rho_0^{(k)})$ the behavior of $\rho(t) \times t$. Only three cases were identified, namely fixed point $\rho = 0$, fixed point $\rho = 1$ and oscillatory behavior of period 2.

In figure 16.2 we can observe that by taking a higher number of runs the phase diagram region corresponding to $\rho = 0$ decreases significantly for $p = 0$ and this is even more pronounced for $p = 1$.

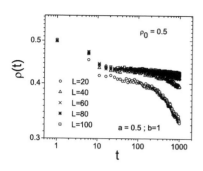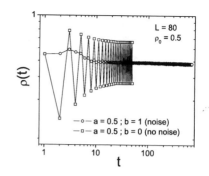

Fig. 16.3. (a) Time evolution of the density of motivated agents for $a = 1/2$, $b = 1$, $\rho_0 = 1/2$, for different lattice sizes ($L = 20, 40, 60, 80, 100$) with $K = 4$. (b) A comparison between a particular case with noise ($b = 1$) and no noise ($b = 0$).

16.5.2 Time Evolution of Bankruptcy-Related Measures in the Small World

We have performed numerical simulations starting with half of the agents motivated ($\rho_0 = 1/2$). These were randomly chosen within the small-world networks, which on their turn were built from lattices with coordination $K = 4$ (i initially connected to nodes $i - 2$, $i - 1$, $i + 1$ and $i + 2$) and periodic boundary conditions. We first measured the evolution of the average density of motivated agents, considering several random initial conditions (different configurations of the small world and of the motivation level of the agents with fixed $\rho_0 = 1/2$). In order to better understand the effect of neighbor density alone we first considered the case $p = 0$.

We also explored the effect of finite size in the density of motivated agents. Our results (see figure 16.3 (a)) for the particular case with noise ($a = 1/2$ and $b = 1$) show a deviation from $\rho(t)$ vs t for small values of L. In that plot, we can also observe a tendency of the density towards a constant value after approximately 20 MC-steps, where $N_s = 1000$ runs were perfomed to compute the average. For $L = 80$ we also simulated the case without noise $a = 1/2$ and $b = 0$. An oscillatory behavior for the density of motivated agents is found, as can be seen in figure 16.3 (b). The noisy and noiseless cases are depicted together. After the 27th MC-step the density oscillates between two fixed values of density ($\rho_1 = 0.29947$ and $\rho_2 = 0.62375$).

We also assessed the possibility of agents going bankrupt. For this we studied four small world configurations: $p = 0, 0.1, 0.2, 0.3$, and measured the average fraction of bankrupt agents, that is

$$f(t) = \frac{1}{L} \#\{W_k(t) < 0\} .$$

where we recall that $W_k(0) = w_0$.

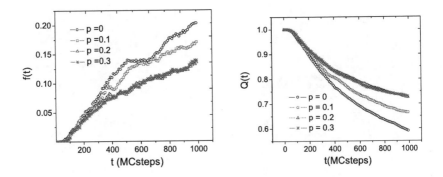

Fig. 16.4. (a) Time evolution of the fraction of agents in a situation of bankruptcy. (b) Probability of a randomly chosen agent not reaching bankruptcy.

For the simulations we considered $w_0 = 10$. A plot of $f(t)$ is shown in figure 16.4 (a). To quantify more precisely the influence of p on the bankruptcy, we used the concept of first return probability $Q(t)$:

$$Q(t) = \frac{1}{L}\#\{W_k(t) < 0\}, \ W_k(t') \geq 0 \quad \text{for all} t' < t.$$

This quantity is the probability that an agent remains wealthy, that is $W_k(t') > 0$ for all $t' < t$. Our results show that this probability decays with a power law and $Q(t)$ becomes less steep with increasing p (figure 4(b)).

16.5.3 Individualist Agents

In the noisy case ($b = 1$ and $a = 1/2$), we have evaluated the functions $f(t)$ and $Q(t)$ for the regular lattice (small world $p = 0$), considering a concept defined in this paper as individualist agent, that means, agents which act without taking heed of their neighbors' action. In this case the agent's decision is based on its proper information: if it is motivated, it invests $S_i = 3$. Otherwise its investment is $S_i = 0$.

Defining the fraction of these players as ϕ, we have performed some tests to determine its effect on the dynamics of $W(t)$, $Q(t)$, $\rho(t)$ and $f(t)$. In figure 16.5 we can observe that an increase of ϕ leads the systems towards ruin. From this one may conclude that it is more satisfactory for the group to comunicate before investing.

16.5.4 Decreasing Odds for Bankruptcy: A Complete Graph

We also analyzed the behavior of $f(t)$ and $Q(t)$ for the noisy case ($a = 1/2$ and $b = 1$) when agents form a complete graph. In this case no agent goes bankrupt, since $Q(t) = 1$ and $f(t) = 0$ for all $t > 0$. This is interesting

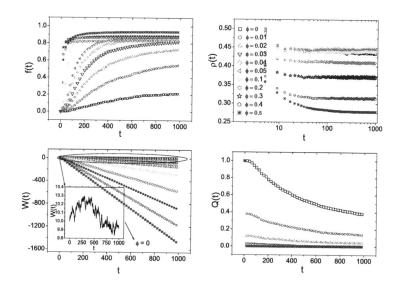

Fig. 16.5. Time evolution of measures $W(t)$, $Q(t)$, $f(t)$ and $\rho(t)$ as a function of ϕ. We considered the cases of $\phi = 0$, 0.01, 0.02, 0.03 ,0.04, 0.05, 0.1, 0.2, 0.3, 0.4, and 0.5. Legends are the same for all figures. The inset plot in the bottom left figure depicts $\phi = 0$, showing that the wealth fluctuates around the initial value of $w_0 = 10$

when compared to the dynamics of the game in small-world networks. The main lesson is: to talk to all agents (which in practice might be impossible) diminishes the risk and makes bankruptcy less probable (since $f(t) = 0$ and $Q(t) = 1$).

It is important to notice that the density of motivated agents goes to 0 in the complete graph as it can be observed in figure 16.6. Thus, basically the investment is blocked. In that figure we plot the wealth and density as a function of time for different sizes of the graph. One may observe that the wealth is practically constant after a given time.

16.6 Conclusions

We explored the emergent dynamics in a modified public investment game, both in small-world networks and in a complete graph, where benefits are determined by two parameters: a deterministic (a) and a random one (b). Investment depends on the motivation level of an agent and of its neighbors. We performed some simulations changing $f(t)$ (fraction of bankrupt agents) and $Q(t)$ (probability that an agent does not go bankrupt up to time t) for

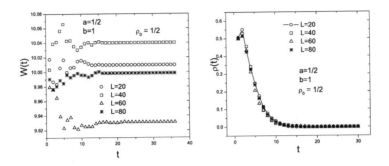

Fig. 16.6. (a) Time evolution of wealth for a public investment game in the complete graph. (b) Time evolution of density in the complete graph.

different small world configurations (parameter p), generated from regular lattices with coordination $K = 4$. Our results show that both $f(t)$ and $Q(t)$ depend on p. For the regular case $(p = 0)$ we studied a noisy $(a = 1/2$ and $b = 1)$ and a noiseless $(a = 1/2$ and $b = 0)$ situation. A finite-size dependence is observed for the case with noise, whereas an oscillatory behavior is observed in the noiseless case for the small world network.

Moreover, a phase diagram can be obtained in a simplified situation $S_i = \sigma_i + \sigma_{i-1}$ for $p = 0$ and $k = 2$, as shown in [7]. Therefore, in this work we have extended the results to obtain the phase diagram for the general case: numerically for small-world networks and as a exact phase diagram for the complete graph.

Our results also corroborate the idea that the probability of an agent going bankrupt is zero even in the presence of fluctuations $(a = 1/2$ and $b = 1)$ if players form a complete graph. This indicates that information exchange is key in the dynamics of the wealth of a group of investors.

References

[1] Abramson G, Kuperman M (2001) Social games in a social network. *Physical Review E.*
[2] Ashlock D (2005) Evolutionary Computation for Modeling and Optimization. Springer, New York
[3] Bazzan ALC & Cavalheiro AP (2003) Influence of Social Attachment in a Small-World Network of Agents Playing the Iterated Prisoner's Dilemma In: Parsons S Gmytrasiewicz P (eds) *5th Workshop of Game Theoretic and Decision Theoretic Agents*, pp. 17–24

[4] Durlauf SN (1999) How can statistical Mechanics contribute to social science? Proc. Natl. Acad. Sci. 96:10582–10584

[5] Milgram S (1967) The small world problem. *Psychol. Today* 2

[6] Nowak M, May R (1992) Evolutionary games and spatial chaos. *Nature* 359:826–829

[7] da Silva R, Bazzan ALC, Baraviera AT, Dahmen SR (2006) Emerging collective behavior and local properties of financial dynamics in a public investment game. To appear in Physica A (2006)

[8] Watts D, Strogatz SH (1998) Collective dynamics of small world networks. Nature 393:440–442

[9] R. Axelrod. *The Evolution of Cooperation*. Basic Books, 1984.

[10] A. L. C. Bazzan, R. Bordini, G. Andriotti, R. Viccari, and J. Wahle. Wayward agents in a commuting scenario (personalities in the minotity game). In *Proc. of the Int. Conf. on Multi–Agent Systems (ICMAS)*. IEEE Computer Science, July 2000.

[11] A. L. C. Bazzan, R. Bordini, and J. Campbell. Moral sentiments in multi–agent systems. In *Intelligent Agents V*, number 1555 in LNAI, pages 113–131. Spriger–Verlag, 1999.

[12] A. L. C. Bazzan and R. H. Bordini. A framework for the simulation of agents with emotions: Report on experiments with the iterated prisoner's dilemma. In J. P. Müller, E. Andre, S. Sen, and C. Frasson, editors, *Proceedings of The Fifth International Conference on Autonomous Agents (Agents 2001), 28 May – 1 June*, pages 292–299, Montreal, Canada, 2001. ACM Press.

[13] S. A. Kauffman. *The Origins of Order*. Oxford University Press, Oxford, 1993.

[14] B. J. Kim, A. Trusina, P. Holme, P. Minnhagen, J. S. Chung, and M. Y. Choi. Dynamic instabilities induced by asymmetric influence: Prisoner's dilemma game in small-world networks. *Physical Review E*, 66, 2002.

[15] S. Milgram. The small world problem. *Psychol. Today*, 2, 1967.

[16] M. Nowak and R. May. Evolutionary games and spatial chaos. *Nature*, 359:826–829, 1992.

[17] M. Ridley. *The Origins of Virtue*. Viking Press, London, 1996. 304 pp.

[18] D. J. Watts and S. H. Strogatz. Collective dynamics of 'small-world' networks. *Nature*, 393(6684):397–498, June 1998.

Social Norms, Cognitive Dissonance and Broadcasting: How to Influence Economic Agents

Andrew Bertie[1], Susan Himmelweit[2], and Andrew Trigg[2]

[1] Open University, Milton Keynes MK7 6AA, UK a.j.bertie@open.ac.uk
[2] Open University, Milton Keynes MK7 6AA, UK s.f.himmelweit@open.ac.uk
[3] Open University, Milton Keynes MK7 6AA, UK a.b.trigg@open.ac.uk

17.1 Introduction

Economists pay little attention to social norms and the processes by which they are formed and change. This omission is regrettable because much economic decision-making is influenced by social norms, and the result of such decision-making may in turn influence social norms. One example is the decision faced by mothers of pre-school children as to whether to remain in employment or to stay at home and care for their children themselves. This decision is subject to many economic factors, crucially the education and thus the earning capacity of the mother, but is also known to be influenced by social norms concerning appropriate ways of caring for pre-school children. It is a decision that in turn has significant economic effects, for example on the ability of European countries to meet employment targets and fund future pensions.

In a study of mothers of pre-school children in the UK Himmelweit and Sigala 2004 found that mothers' attitudes were influenced by both the attitudes of those around them and by their own experience. In particular, where their behaviour was in conflict with their own attitudes, mothers were more likely to change one or other of these, than when their behaviour and attitudes were consistent with each other. They also found a change in attitude to be more likely than behavioural change. These findings are consistent with the psychological theory of cognitive dissonance, that people experience discomfort from a poor fit between attitudes and behaviour. Cognitive dissonance can be resolved by changing behaviour, but it is at least as frequently resolved by changing attitudes (Festinger 1957, Pungello and Kurtz-Costes 2000).

Mothers' attitudes and behaviour changed not only as a result of cognitive dissonance but also through social influence. It was found that mothers were more likely to interact with, and therefore be influenced in their attitudes by, others who behaved similarly to themselves. This suggests that it would be useful to build a model that combines such individual and social influences. Such a model would be applicable to the wide class of situations in which social norms influence decision-making not only directly, but also indirectly through internalization of such norms into personal attitudes.

Several agent-based models of social influence have been developed. Gilbert and Troitzsch 2005 discuss particularly two models: that of Latané et al. 1994 who use a simple plausible formula to calculate the "persuasive impact" of local neighbours on an individual's attitude, and Schelling's 1971 even simpler segregation model that bases a decision to move on the proportion of immediate neighbours who are ethnically different to an individual. These simple models of social influence give interesting emergent effects such as clustering. Using a "majority rule heuristic" similar to Schelling's model, Miller and Page 2004 discuss various models of social influence in which agents respond in a binary manner to the strength of signals from other agents, which may be in conflict with their own beliefs (functioning like cognitive dissonance in our model). Picker 1997 also uses agent-based modelling to simulate the emergence of social norms.

A powerful and well-known agent-based model developed by Axelrod 1997 shows how culturally homogeneous groups can form by modelling the transmission of cultural features that are subject to social influence. Individual agents, located in a two-dimensional grid, hold a vector of such "features", each of which can take a number of different categorical values, known as "traits". A "social influence event" is the random selection of an agent on the grid to be the "active" cell, then one of its neighbours also at random. The probability of interaction with the selected neighbour depends on the number of the features they have in common. If an interaction occurs, the active agent changes one of the features in which it differs from its neighbour to match the neighbour's.

This simple model has proved very rich in its implications. Axelrod shows that starting from a random distribution of traits among agents, given time a number of cultural regions consisting of agents with identical features will result. In equilibrium, when no further interactions can take place, the agents of such cultural regions have no features in common with those of neighbouring regions. As expected, the equilibrium number of regions was found to increase with the number of traits per feature. However, the number of regions in equilibrium also fell with the size of the grid and with the number of features. These results were less intuitive but could be explained by the power of the social mechanism itself; both larger regions and a larger number of features give more scope for social influence to make agents more similar to each other, even if the range of possible difference was also increased.

17.2 Cognitive Dissonance

This paper considers the results of introducing a new type of event to Axelrod's basic model that represents an internal resolution of the cognitive dissonance experienced when attitudes are not in line with behaviour. In Axelrod's basic model 100 agents living on a 10 x 10 grid have five unidentified cultural features with 10 possible trait values for each feature. In our model the first two features are only binary and identified respectively as a particular form of behaviour (such as 0 = "employed" or 1 = "at home looking after family") and a related attitude (such as 0 = "pro-employment for mothers" and 1 = "pro-mothers staying at home"). In addition each agent has three further features, representing other identified cultural characteristics that play a part in social influence, each of which can take 10 trait values as in Axelrod's basic model. The model was implemented in the *Repast* agent-based modelling system (North et al. 2006) with the random selection of an agent as an active cell for a potential "social influence event" occurring at each *Repast* tick.

As expected from Axelrod's results, binary features reduce the number of regions in equilibrium considerably. In fact, as Axelrod showed, if all features are binary there can be at most two final regions, but in practice such equilibria are rare and a monoculture of only one region is by far the most likely result when they are binary features. Running an Axelrod type model but with only two features binary and the remaining three taking 10 trait values, convergence to a monoculture happened in 99.9% of cases.

17.2.1 Symmetric Dissonance Events

The innovation in our model is that there is an internal connection between the first two features. If their trait values are not equal (i.e. at home looking after family but pro-employment, or employed but pro-mothers staying at home) then the agent is in a state of cognitive dissonance. An agent in this state if selected for a "cognitive dissonance event" will change a trait value to make the values of its first two features match. These internal "cognitive dissonance events" are scheduled separately from the usual external Axelrod social influence events between agents and occur with probability p at each *Repast* tick on a randomly chosen agent, where p is a "cognitive dissonance probability" parameter.

To investigate the effects of cognitive dissonance events, this model was run 500 times for each value of $p = 0.2, 0.4, 0.6, 0.8$ and 1, and 1000 times for $p = 0$, no cognitive dissonance events. Table 17.1 shows that the number of equilibrium regions was predominantly 1 for all values of p, including 0, with smaller frequencies of 2 to 8 regions. Multiple regions are slightly more likely at higher values of p.

There is some evidence, therefore, that cognitive dissonance events reduce the likelihood of complete cultural uniformity, but this effect is quite weak and not strongly affected by the probability of these events. It happens because

Table 17.1. Symmetric dissonance events: number of stable regions in equilibrium from 500 runs each at 5 positive values of the cognitive dissonance event parameter, p, and 1000 runs for $p=0$ (no cognitive dissonance events)

	0	.20	.40	.60	.80	1.00
1	999	484	465	467	456	458
2	1	12	27	20	30	29
3		4	4	6	8	6
4			3	3	4	5
5				2		1
6				2		
7			1			
8					2	1

cognitive dissonance events can in a few cases induce faster convergence to an equilibrium that does not leave enough time to reach uniformity. The slower process of cultural convergence without cognitive dissonance events leaves enough time that, in all but one case above, convergence to a uniform monoculture takes place.

This can be illustrated by comparing the time series plots of two different runs with the same cognitive dissonance parameter that converge at different rates. In Figure 17.1a convergence to 8 final regions was more rapid than in Figure 17.1b which converged to a single region (note different scale). Also in Figure 17.1a all cognitive dissonance was eliminated before complete convergence to a multi-region equilibrium, while in Figure 17.1b convergence to a monoculture occurred simultaneously with the elimination of cognitive dissonance.

Figure 17.2 shows the grid of agents for the exceptional case of 8 final regions occurring at $p = 0.8$, corresponding to Figure 17.1a. The circles represent mothers at home (32%) and rectangles represent employed mothers (68%). In general, however, the simulations converged to a monoculture of either 100% mothers at home or 100% employed mothers, with equal likelihood.

17.2.2 Asymmetric Dissonance Events

So far we have assumed that cognitive dissonance events are symmetrical with regard to behaviour and attitude, that they are equally likely to involve a change in behaviour or a change in attitude. However, economists (since they like to assume fixed preferences) are more likely to assume people change their behaviour rather than their attitudes, while the psychological theory of cognitive dissonance suggests that attitude change may be more likely. So it seems worthwhile to investigate the effect of an "asymmetric dissonance parameter" q that controls the direction of dissonant events, and is the probability that a cognitive dissonance event is a change of the agent's attitude rather than a change in the agent's behaviour. A value of $q = 0.5$ therefore produces the

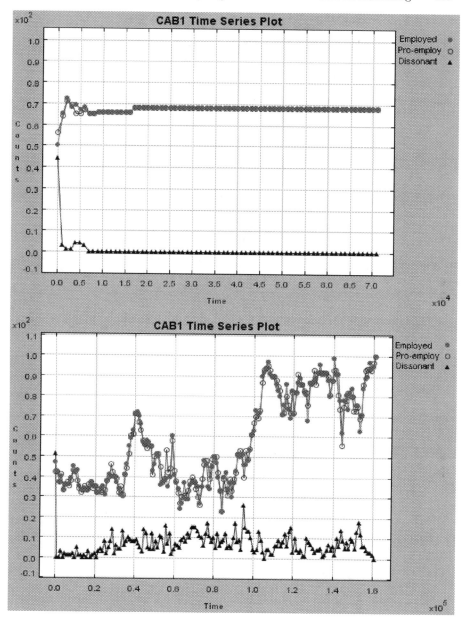

Fig. 17.1. Time series plots of numbers of agents that are *employed*, *pro-employment* and *dissonant* for
$p = 0.8$, (a) converged to 8 final regions (b) converged to 1 final region

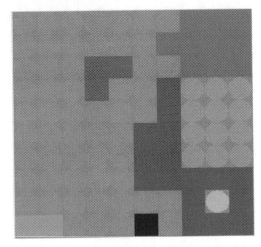

Fig. 17.2. Grid display with 8 final regions at $p = 0.8$

symmetric dissonance events discussed above, $q = 1$ produces an event that always changes attitude, and $q = 0$ produces an event that always changes behaviour.

Table 17.2 shows that for extreme anti-symmetry, i.e. cognitive dissonance events that are only behaviour-changing or only attitude-changing, a larger, though still small, proportion of the simulations did not converge to a mono-culture than was the case for symmetric dissonance (Table 17.2 shows runs just for $q = 0$, but $q = 1$ should produce similar results by symmetry). A possible explanation for this is that asymmetric dissonance leads to a faster elimination of dissonance and therefore on more occasions to convergence before complete uniformity has been achieved on other features.

Table 17.2. Asymmetric dissonance events: number of stable regions in equilibrium from 500 runs each at 5 positive values of the cognitive dissonance event parameter, p, with asymmetric dissonance parameter $q = 0$ and 1000 runs for $p= 0$ (no cognitive dissonance events)

	0	.20	.40	.60	.80	1.00
1	999	456	457	457	450	440
2	1	36	32	23	31	32
3		6	9	13	13	24
4		1		4	2	1
5				1	2	2
6			2	2	2	1
7		1				
8						

17.3 Broadcasting

In his original paper Axelrod explores the effects of cultural drift, random mutations of the traits of agents. These have the effect of disrupting convergence to a multi-region equilibrium, by allowing further social interaction events, or even preventing convergence altogether. Klemm et al. 2005 uses the term exogenous perturbations to describe this cultural drift mechanism. Cognitive dissonance events could be described as endogenous perturbations, which work in the opposite direction, aiding convergence and making multi-region convergence more likely as Table 17.1 showed.

Broadcasting can be modelled as another form of exogenous perturbation, originally suggested by Axelrod as an external agent with fixed traits interacting at random with agents in the grid. To explore broadcasting as a tool of policy, let us suppose there is a government keen on encouraging mothers into employment. It has two alternative broadcast strategies. One is to send out the same consistent pro-employment message repeatedly, using a constant set of traits for the remaining three features, including "employed" for the behaviour feature. This could be thought of as a series of fixed broadcasts using an identical image of a working mother expounding a pro-employment attitude. The other strategy is to send out random broadcasts that are constant in expounding a pro-employment attitude, but show mothers whose other four features vary randomly (including whether the mother doing the broadcast is herself in employment or not).

17.3.1 Fixed Broadcasts

The effect of introducing fixed broadcasts is greatly to increase the number of final regions, to such an extent as to completely eliminate monocultures. Increasing the rate of broadcasting increases the number of final regions. These effects are counterintuitive: broadcasting could be expected to drive the agents towards a monoculture; increasing the broadcast rate might be thought to make a monoculture more likely. These counterintuitive results may be explained by a strongly transmitted culture overriding possibilities of sites changing their traits in other directions towards convergence. In other words, high-rate global broadcasts of fixed messages promote local heterogeneity. For example, Figure 17.3 shows the effects of a fixed pro-employment broadcast on the number of final regions with various levels of symmetric cognitive dissonance events ($q = 0.5$). Recall from Table 17.2 that without broadcasting the median number of regions in equilibrium was 1 for all levels of p. Indeed without broadcasting 90% of all runs converged to a monoculture except when $p= 1.0$, and even then 60 out of 500 did so. With broadcasting, higher rates of p again increase the number of final regions, but at a much higher level than before.

Another interesting and counterintuitive result concerns the efficacy of broadcasts. Figure 17.4 plots the percentage of agents in employment for the

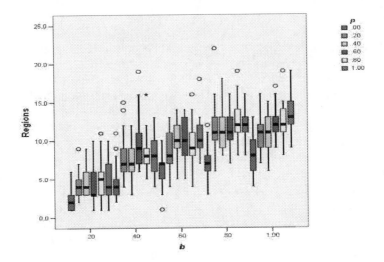

Fig. 17.3. Clustered boxplots of number of final regions at 5 levels of b for a constant pro-employment broadcast and 6 levels of p (25 runs per boxplot), for symmetric cognitive dissonance events

runs that produced Figure 17.3. From this we can see that broadcasting has the desired effect and employed mothers form a majority in all cases. (Recall from Table 17.2 that without broadcasting the equilibrium position is a mono-culture in the vast majority of cases, either 100% mothers employed or 100% at home with family, with the average percentage of mothers in employment being 50%.) So broadcasting increases the equilibrium percentage employed on average and makes it more predictable. Counterintuitively, however, this percentage is lower at higher broadcast rates – probably for the same reason as before, that high-rate broadcasts promote faster convergence and hence local heterogeneity. While the majority culture is identical to the broadcast culture, pockets of alternative cultures survive in equilibrium that must differ from the majority culture in all features.

Looking at the interaction with p, the cognitive dissonance parameter, Figure 17.4 suggests that cognitive dissonance events dilute the effect of broadcasting. As Figure 17.3 showed, increasing levels of cognitive dissonance increase the number of regions with alternative cultures, but Figure 17.4 shows that these regions are then in total smaller and the majority culture more dominant. (There appears to be little systematic difference between the effects of symmetric and asymmetric dissonance events on the efficacy of broadcasting, so only the case of symmetric dissonance events is shown. If anything, behaviour-changing cognitive dissonance events ($q = 0$) are slightly less disruptive and attitude-changing cognitive dissonance events ($q = 1$) slightly more disruptive of the effects of broadcasting on the percentage employed.)

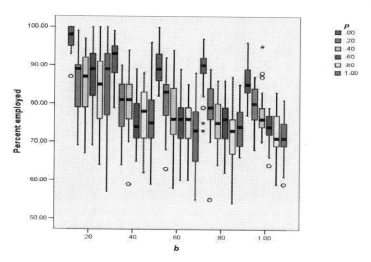

Fig. 17.4. Clustered boxplots of percent employed at 5 levels of b for a constant pro-employment broadcast and 6 levels of p (25 runs per boxplot), for symmetric cognitive dissonance events

Figure 17.5 shows a fairly typical final agent grid resulting from $b = p = 1$. The pro-employment culture has produced a prevailing culture of employed agents (black squares) whose culture matches the broadcast culture in every feature, while isolating 9 regions of agents not in employment whose cultures are diverse and have no features in common with the dominant monoculture.

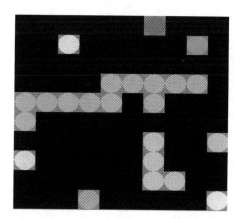

Fig. 17.5. Grid display with 10 final regions and 23% carers: $b = 1$ (pro-employment), $p = 1$

17.3.2 Random Broadcasts

Random broadcasts, like a fast enough constant stream of exogenous perturbations, prevent the model ever converging. They are by themselves rather
ineffective in promoting the behaviour they want to encourage. Figure 17.6
shows the effect on the percentage of mothers employed at different levels
of intensity of a random pro-employment broadcast if there are no cognitive
dissonance events. Random pro-employment broadcasts, that is broadcasts
that are random in all features except their pro-employment attitude, do not
on average produce a higher level of employment. (Recall from Table 17.1
that without broadcasting or cognitive dissonance events convergence to a
monoculture happens in 99.9% of cases – so although the median for $b = 0$
in Figure 17.6 is 100% this is just chance and in almost as large a number
of cases there is convergence to a monoculture where the number of mothers
employed is 0%, with the average number of employed mothers across all cases
being 50%.) This is not surprising because without any specific connection between attitudes and behaviour, however effective broadcasts are at changing
attitudes they will have no specific effect on changing behaviour in a related
direction. However, Figure 17.6 shows that random broadcasting does have
an effect on the number of regions so that monocultures are eliminated and
in all equilibria cultures with mothers in employment coexist with cultures in
which mothers are at home looking after their families.

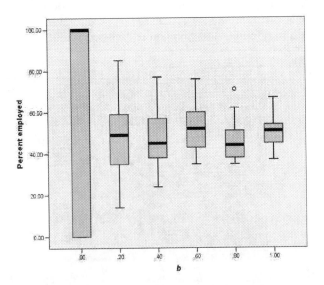

Fig. 17.6. Clustered boxplots of percent employed at 6 levels of b for a random
pro-employment broadcast with no cognitive dissonance events ($p = 0$)

If there is cognitive dissonance, however, Figure 17.7 shows that the effectiveness of random broadcasts varies considerably according to the type of cognitive dissonance events (recall there was little such difference in the case of fixed broadcasts). Even low levels of behaviour-changing dissonance events can turn random broadcasts into highly effective tools of behavioural change, more effective than fixed broadcasts, and higher levels of such cognitive dissonance events can make random broadcasts so effective that they produce median levels of employment of almost 100%. As with fixed broadcasts, less frequent random broadcasts are more effective than more frequent ones, particularly when there is a high rate of behaviour-changing dissonance events.

The effects when there are attitude-changing dissonance events could not be more different. Attitude-changing dissonance events make random broadcasts worse than useless in changing behaviour, reducing the mean numbers employed to well below 50%, and higher levels of such dissonance events depress the numbers of employed mothers yet further. More frequent broadcasts in this case reduce the counterproductive effect of having the broadcasts at all, a result similarly contradictory to that found in the case of behaviour-changing dissonance events where broadcasts are highly effective but a higher rate of broadcasts less effective than a lower one.

Symmetric broadcasts have an intermediate effect, raising the median rate of employment in all cases. But there is no discernable pattern as to whether increasing the overall rate of broadcasting or whether higher or lower rates of cognitive dissonance events makes broadcasting more or less effective. In this case, random broadcasts are more effective than nothing but in general less effective than fixed broadcasts (see Figure 17.4).

These rather startling differences can be explained perhaps by remembering that random broadcasts are constant only in their attitude feature. Fixed broadcasts can affect only those agents who already have some feature in common with the transmitted culture. As Figure 17.3 showed, fixed broadcasts therefore tend to leave pockets of cultures completely at variance with the dominant culture they promote. Random broadcasts, because their non-attitude features vary to take on the whole range of possible values, can influence any agent and therefore can be effective in changing any agent's attitude.

However, unlike fixed broadcasts that have an equal impact on the behaviour and attitudes of the agents they affect, the only change random broadcasts systematically promote is in attitudes. Random broadcasts therefore need behaviour-changing dissonance events to make a systematic impact on behaviour. When there are such events, random broadcasts are highly effective in promoting employment. Because random broadcasts reach everyone they are more effective than fixed broadcasts when the only dissonance events are behaviour-changing. In the symmetric case ($q = 0.5$) behaviour-changing dissonance events still occur, but there are also counteracting attitude-changing dissonance events.

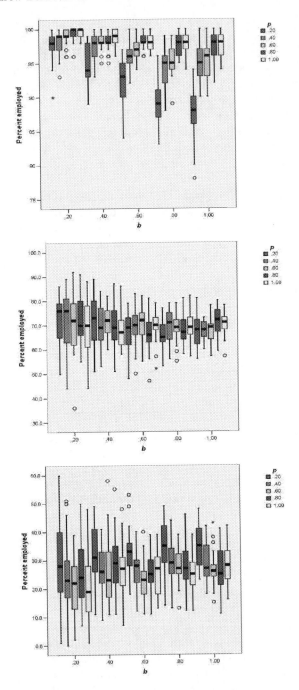

Fig. 17.7. Clustered boxplots of percent employed at 5 levels of b for random pro-employment broadcasts and 5 levels of p (25 runs per boxplot) for (a) behaviour-changing cognitive dissonance events ($q = 0$) and (b) symmetric cognitive dissonance events ($q = 0.5$) and (c) attitude-changing cognitive dissonance events ($q = 1.0$)

The puzzle then is why attitude-changing cognitive events reduce the numbers of employed mothers in equilibrium. They have this effect both in the symmetric case when compared with the case with behaviour-changing dissonance events alone ($q = 1$), and most spectacularly in the case of only attitude-changing cognitive behaviour ($q = 0$) when random broadcasting becomes counterproductive. The reason must be something to do with the way these dissonance events dilute the effects of broadcasting by encouraging faster convergence to more diverse regions. However it not obvious why this should work to such an extent as to reduce the numbers of mothers employed.

17.4 Dynamism, Polarization and Equilibria

Axelrod's analysis of his original model focused on convergence to equilibria. Figure 17.1 shows that the convergence to the simplest equilibrium (one region) can be slow and complex with large shifts in the numbers of employed, whereas convergence to a more complex equilibrium (eight regions) can be fast and relatively uninteresting. Exogenous perturbations (broadcasting or cultural drift) or endogenous perturbations (cognitive dissonance events) further compromise convergence to complete equilibrium, though they may be important factors which produce interesting effects.

These observations lead us to consider measures of dynamism, polarization and "degrees of equilibria" that are less orientated to the concept of a final equilibrium. Latané et al. 1994, a paper that seems to have been largely neglected by agent-based modellers, discuss such measures in the context of social attitudes and behaviour. (We do not consider their suggested measures of clustering, since we use Axelrod's method of counting cultural regions.)

17.4.1 Dynamism

The dynamism of individual simulations can be assessed subjectively from time series plots such as those in Figure 17.1, but to analyze many simulations and draw objective conclusions requires formal measures of dynamism. Following Latané et al., we measure the dynamism of our models by the average number of changes of attitude per agent until convergence or within the first 100,000 *Repast* 'ticks', whichever happens first. We investigated the dynamism of our model by running 25 simulations at 6 levels of dissonance event probabilities, p, 6 levels of broadcast probabilities, b, 3 levels of the asymmetric dissonance parameter, $q = 0$, 0.5 & 1, and for fixed and random broadcasts.

Consider first fixed broadcasts, which always converge. Table 17.3 shows the average number of changes in attitude before convergence with fixed broadcasts at different levels of broadcasting, and symmetric cognitive dissonance events ($q = 0.5$). A table for dynamism of behaviour would show a similar pattern of values, as would tables for $q = 0$ and $q = 1$ with fixed broadcasts, for dynamism of both attitude and behaviour.

Clearly broadcasting decreases dynamism and increasing rates of broadcasts decrease it further. Broadcasting may produce more complex equilibria (see Figure 17.4, but less dynamic paths to those equilibria. Table 17.3 also shows that when there is broadcasting the effect of increasing the rate of dissonance events is to reduce dynamism in the model. This is consistent with the earlier remark that cognitive dissonance events can induce faster convergence to equilibrium – leaving less time for attitudes to change, although it does not seem to apply consistently when $b = 0$, perhaps because in these cases, as Table 17.1 showed, convergence to a monoculture remains by far the most likely outcome, even at high levels of p.

Table 17.3. Attitude dynamism (mean of 25 runs per cell) for 6 values of b (rows) and 6 values of p (columns), fixed broadcasts and symmetric dissonance events.

b	p					
---	0.0	0.2	0.4	0.6	0.8	1.0
0.0	24.5	30.28	30.54	34.79	19.88	24.85
0.2	2.90	2.00	2.31	1.67	2.19	2.40
0.4	1.56	1.47	1.34	1.00	.99	.97
0.6	1.03	1.04	.86	.76	.65	.63
0.8	.98	.75	.58	.62	.51	.52
1.0	.80	.69	.59	.57	.51	.43

Where convergence does not occur, the level of dynamism measures the average number of changes of attitudes and behaviour over a period (100,000 *Repast* ticks) that is longer than convergence usually takes. So, unsurprisingly, the level of dynamism measured as a result of random broadcasts which do not converge is greater than for fixed broadcasts which do. With random broadcasts levels of dynamism differ more between attitude and behaviour, and across different levels of q.

Tables 17.4 and 17.5 show average dynamism of attitude with random broadcasts in the cases of behaviour-changing dissonance events ($q = 0$) and attitude-changing dissonance events ($q = 1$). Table 17.4 shows that if there are only behaviour-changing dissonance events broadcasting suppresses dynamism of attitude for any value of p and increasing rates of broadcasting events suppress it further. Cognitive dissonance events reduce dynamism throughout. This must be by reducing opportunities for attitude change through interaction with neighbouring sites or broadcasts. That random broadcasting decreases dynamism in this case must be through it effectively creating a uniformity of attitude and thus reducing opportunities for attitude change thorough interaction with neighbouring sites.

Table 17.4. Attitude dynamism (mean of 25 runs per cell) for 6 values of b (rows) and 6 values of p (columns), random broadcasts and behaviour-changing dissonance events.

b	0.0	0.2	0.4	p 0.6	0.8	1.0
0.0	30.85	25.39	16.11	18.32	19.63	16.55
0.2	5.79	3.70	3.00	2.87	2.57	2.33
0.4	3.26	2.06	1.87	1.63	1.59	1.30
0.6	2.22	1.48	1.40	1.24	1.15	1.19
0.8	1.48	1.29	1.17	1.14	1.02	1.06
1.0	1.40	1.17	1.04	.91	.87	.86

Table 17.5 shows average attitude dynamism when there are only attitude-changing dissonance events. When $p = 0$ there are no cognitive dissonance events and random broadcasting decreases dynamism and increasing rates of broadcasts decrease it further (this is the same situation as in the previous table in the absence of cognitive dissonance events; though the somewhat different numbers reflect different runs). With attitude-changing cognitive dissonance events, dynamism is greater than without them, though higher rates of cognitive dissonance reduce dynamism and increased rates of broadcasts increase dynamism. It must be that dissonance events again reduce opportunities for attitude change through interaction with neighbouring sites. But an indirect opportunity for attitude change arises in this case through attitude-changing cognitive dissonance events consequent upon behavioural changes brought about by random broadcasts. The difference in attitude dynamics in Tables 17.4 and 17.5 may go some way towards explaining why random broadcasts are so effective when cognitive dissonance events are behaviour-changing and so ineffective, indeed counterproductive, when they are attitude-changing.

Table 17.5. Attitude dynamism (mean of 25 runs per cell) for 6 values of b (rows) and 6 values of p (columns), random broadcasts and attitude-changing dissonance events.

b	0.0	0.2	0.4	p 0.6	0.8	1.0
0.0	25.56	23.22	18.39	19.54	15.85	12.91
0.2	5.65	50.99	41.26	35.12	32.18	29.81
0.4	2.58	62.68	52.99	47.61	44.09	42.17
0.6	1.97	68.83	63.34	58.50	54.29	53.06
0.8	1.56	72.86	71.61	67.85	65.34	62.97
1.0	1.50	76.19	78.69	75.84	75.13	72.58

17.4.2 Degrees of Equilibria

Latané et al. also discuss what they call *degrees of equilibria*, which they classify into four types: *unification, stable equilibria, dynamic equilibria* and *disordered evolution*. Unification occurs when the agents reach a consensus and converge on a common attitude. We have seen that this is a common outcome of our models. Stable equilibria are identified as incompletely polarized systems in which further interaction no longer leads to change. This describes Axelrod's convergence to more than one stable region, which our models are less likely to achieve without internal or external perturbations. Dynamic equilibria represent cases where the system achieves global order in terms of polarization and clustering at an intermediate level, but where individual agents continue to change. We have also observed this in our models by viewing the grid of agents in individual simulations. Disordered evolution describes cases that exhibit continued change at the agent level without the emergence of any global order. Such cases are characterized by high dynamism, many cultural regions, and polarization that depend on the initial state of the model. We have not observed this in our models. Even where there is disequilibrium, there is some evidence of order at the agent level when we view the grid of agents. We also do not find it associated with high dynamism.

17.5 Conclusion

This paper illustrates some of the issues that can be discussed using a model of social influence that also incorporates cognitive dissonance and broadcasting as internal and external perturbations, respectively. Cognitive dissonance has been modelled as a process by which agents resolve a bad fit between their attitudes and their behaviour by changing one or other of these. Broadcasting has been modelled as the creation of an external agent interacting randomly with any of the agents on the grid. This paper has shown that adding processes of cognitive dissonance to Axelrod's model of social influence speeds up convergence in some cases, making the survival of cultural heterogeneity more likely. Broadcasts have been shown to influence behaviour, but occasionally in an unintended direction and they are generally less effective if repeated too frequently. Broadcasts that convey a fixed message are effective in promoting the intended behaviour but are less effective the higher the rate of cognitive dissonance events. Broadcasts that are random in features other than an attitude promoting the intended behaviour can be more or less effective than fixed broadcasts depending on the prevailing type of cognitive dissonance events. These different types of broadcast vary in their effectiveness at least in part because they result in different levels of dynamism of attitudes and behaviour.

This suggests that a government interested in using broadcasting to change behaviour, to promote employment among mothers of young children for example, would do well to study the prevailing modes of cognitive dissonance.

If the prevailing mode is attitude-changing, that is that faced with a bad fit between attitudes and behaviour people tend to change their attitude, the government would achieve its aims most effectively through a not too frequently repeated fixed broadcast that promotes both the desired behaviour and the attitude that supports it. If, on the other hand, the dominant mode of cognitive dissonance resolution is for people to change their behaviour, the government would do better with a random broadcast that promotes a pro-employment attitude but through a variety of different images that speaks to agents across the whole range of cultures. Provided these broadcasts are not repeated too often they can be spectacularly effective, promoting nearly universal conformity to the behaviour the government wishes to encourage. If there are both types of reactions to cognitive dissonance, the best procedure to follow depends on the balance between the two; at equal rates of the two types of cognitive dissonance events fixed broadcasts are somewhat more effective than random ones.

References

[1] Axelrod R (1997) The dissemination of culture: a model with local convergence and global polarization. J Conflict Resolution 41: 203–226
[2] Gilbert N and Troitzsch KG (2005) Simulation for the Social Scientist. Open University Press
[3] Himmelweit S, Sigala M (2004) Choice and the relationship between identities and behaviour for mothers with pre-school children: some implications for policy from a UK study. J Social Policy 33: 455–478
[4] Festinger L (1957) A Theory of Cognitive Dissonance. Stanford University Press, Stanford
[5] Klemm K, Eguíluz VM, Toral R, San Miguel M (2005) Globalization, polarization and cultural drift. J Economic Dynamics & Control 29: 321–334
[6] Latané B, Nowak A, Liu JH (1994) Measuring emergent social phenomena: dynamism, polarization, and clustering as order parameters of social systems. Behavioural Science 39: 1–24
[7] Miller JH and Page SE (2004) The standing ovation problem. Complexity 9:8–16
[8] North MJ, Collier NT, Vos JR (2006) Experiences creating three implementations of the Repast agent modelling toolkit. ACM Transactions on Modelling and Computer Simulation 16:1–25
[9] Picker RC (1997) Simple games in a complex world: a generative approach to the adoption of norms. University of Chicago Law Review 64:1225–1288
[10] Pungello, EP and Kurtz-Costes B (2000) Working women's selection of care for their infants: a prospective study. Family Relations 49:245–256

[11] Schelling TC (1971) Dynamic models of segregation. J. Mathematical
     Sociology 1:143–186

Part VI

Methodological Issues and Their Application

Confronting Agent-Based Models with Data: Methodological Issues and Open Problems

Giorgio Fagiolo[1], Alessio Moneta[2], and Paul Windrum[3]

[1] Faculty of Economics, University of Verona (Italy), and Laboratory of
Economics and Management, Sant'Anna School of Advanced Studies, Pisa
(Italy) giorgio.fagiolo@sssup.it
[2] Max Planck Institute of Economics, Evolutionary Economics Group, Jena
(Germany) moneta@econ.mpg.de
[3] Manchester Metropolitan University Business School, Manchester (UK) and
MERIT, University of Maastricht (The Netherlands) p.windrum@mmu.ac.uk

Summary. This paper addresses the problem of finding the appropriate method
for conducting empirical validation in AB models. We identify a first set of issues
that are common to both AB and neoclassical modellers and a second set of issues which are specific to AB modellers. Then, we critically appraise the extent to
which alternative approaches deal with these issues. In particular, we examine three
important approaches to validation that have been developed in AB economics: indirect calibration, the Werker-Brenner approach, and the history-friendly approach.
Finally, we discuss a set of open questions within empirical validation.

18.1 Introduction

Agent-based (AB) researchers in economics have enjoyed significant success
over the last twenty years. The models that have been developed indicate
the viability and vitality of an alternative to mainstream neoclassical economics. Indeed, deep philosophical differences exist between neoclassical and
AB modellers regarding the world faced by real-world agents and, hence, the
type of models that it is useful for economists to construct. AB modellers
reject the aprioristic commitment of new classical models to individual hyper-
rationality, continuous equilibrium, and representative agents. Everything in
the neoclassical world can, in principle, be known and understood. It is often
assumed that the entire set of objects in the world (e.g. techniques of production, or products) is known at the outset. The opposite is the case in the AB
world. Here the set is unknown, and agents must engage in an open-ended
search for new objects. Associated with this distinction are important differences with regards to the types of innovative learning and adaptation that are
considered, definitions of bounded rationality, the treatment of heterogeneity amongst individual agents and the interaction between these individuals,

and whether the economic system is characterized as being in equilibrium or far from equilibrium. Mainstream economists have often recognized the significance of the AB *Weltanschauung*, and have reacted by extending their own modelling framework to incorporate (certain) aspects of heterogeneity, bounded rationality, learning, increasing returns, and technological change. Another sign of the vitality of the AB community has been the development of its own specialist international journals and annual conferences, and the diffusion of its ideas to other areas such as management science, political science and to policy circles.

Nevertheless, there is a perceived lack of robustness in AB modelling, due to the problematic relationship between AB models and empirical data. There is a lack of standard techniques not only for constructing and analyzing AB models, but also to conduct empirical validation. Key areas of debate include: is a 'realist' methodology appropriate? Why should empirical validation be the primary basis for accepting or rejecting a model? Do other tests of model validation exist than the reproduction of stylised facts? If we do proceed down the path of empirical validation, then how should one relate and calibrate the construction of parameters, initial conditions, and stochastic variability in AB models to the existing empirical data? Which classes of empirically observed objects do we actually want to replicate? How dependable are the micro and macro stylised facts to be replicated? To what extent can we truly consider output traces to be stylised facts or, alternatively, counterfactuals? What are the consequences, for the explanative power of a model, if the stylised facts are actually 'unconditional objects' that only indicate properties of stationary distributions and, hence, do not provide information on the dynamics of the stochastic processes that generated them?

The aim of this paper is to provide a critical overview of how AB modellers have been tackling the issue of empirical validation. A strongly heterogeneous set of approaches can be found in the AB literature. An important (and novel) contribution of the paper is a taxonomy that maps the different dimensions of the empirical validation approaches found in AB models. In the next section we shall draw attention to some crucial issues of empirical validation, faced by both AB and neoclassical modellers.

18.2 Core Issues of Empirical Validation

Any model isolates some features of an actual phenomenon. It is usually assumed, in economics as in any other science, that some causal mechanism (deterministic or non-deterministic) has produced the data. We call this causal mechanism "real-world data generating process" (rwDGP). A model approximates portions of the rwDGP by means of a "model data generating process" (mDGP). The mDGP must be simpler than the rwDGP and generates a set of simulated outputs. The extent to which the mDGP is a good representation of the rwDGP is evaluated by comparing the simulated outputs of the mDGP with the real-world observations of the rwDGP. We identify a set of key methodological issues associated with this process of backward induction. These issues are generic in empirical validation, and so apply to neoclassical and AB economists alike.

The first issue is how to deal with the trade-off between concretisation and isolation. Faced with the essential complexity of the world, scientific (not only economic) models proceed by simplifying and focusing on the relationships between a very limited number of variables. Is it possible to model all the different elements of the rwDGP? How can we possibly know all the different elements of the rwDGP? Leading economists (for example, J.S. Mill and J. M. Keynes) have in the past expressed serious doubts about whether we can expect to have models that are fully concretised. In a highly complex world, a fully concretised model would be a one-to-one mapping of the world itself! Thus, economists usually agree that models should isolate some causal mechanisms, by abstracting from certain entities that may have an impact on the phenomenon under examination [13]. A series of open questions remains. How can we assess that the mechanisms isolated by the model resemble the mechanisms operating in the world? In order to isolate the mechanisms, can we make assumptions 'contrary to fact,' that is, assumptions that contradict the knowledge we have of the situation under discussion? This also related to the trade-off between analytical tractability and descriptive accuracy that is faced by all theoreticians seeking to model markets and other economic systems. Indeed, the more accurate and consistent is our knowledge about reality with respect to assumptions, and the more numerous the number of parameters in a model, the higher is the risk of failing to analytically solve the model. By contrast, the more abstract and simplified the model, the more analytically tractable it is. The neoclassical paradigm comes down strongly on the side of analytical tractability.

This brings us to the second core issue of empirical validation: instrumentalism versus realism. Realism, roughly speaking claims that theoretical entities 'exist in the reality,' independent of the act of inquiry, representation or measurement [14]. On the contrary, instrumentalism maintains that theoretical entities are solely instruments for predictions and not true descriptions of the world. A radical instrumentalist is not much concerned with issues of empirical validation, in the sense that (s)he is not much interested in making the model resemble mechanisms operating in the world. His/her sole goal is prediction. Indeed, a (consistent) instrumentalist is usually more willing than a realist to 'play' with the assumptions and parameters of the model in order to get better predictions. While the neoclassical paradigm has sometimes endorsed instrumentalist statements à la Friedman [7], it has never allowed a vast range of assumption adjustments in order to get better predictions. In this sense it fails to be consistent with its instrumentalist background.

The third issue is related to the choice of a pluralist or apriorist methodology. Methodological pluralism claims that the complexity of the subject studied by economics and the boundedness of our scientific representations implies the possibility of different levels of analysis, different kinds of assumptions to be used in model-building, and legitimacy of different methodological positions. Apriorism is a commitment to a set of a priori assumptions. A certain degree of commitment to a set of priori assumptions is normal in science. Often these assumptions correspond to what Lakatos [9] called the 'hard core' assumptions of a research program. But strong apriorism is the commitment to a set of a priori (possibly contrary to the facts) assumptions that are never exposed to empirical validation (e.g. general equilibrium and perfect rationality). Theory is considered prior to data and it is denied the possibility of interpreting data without theoretical presuppositions. Typically, strong apriorist positions do not allow a model to be changed in the face of anomalies, and encourages the researcher to produce ad hoc excuses whenever a refutation is encountered. Lakatos [9] dubbed the research programs involved with such positions as 'degenerating.'

The fourth issue regards the under-determination or identification problem. This is the problem that different models can be consistent with the data that is used for empirical validation. The issue is known in the philosophy of science as the 'under-determination of theory by data.' In econometrics, the same idea has been formalised and labelled 'the problem of identification.' As Haavelmo [8] noted, it is impossible for statistical inference to decide between hypotheses that are observationally equivalent. He suggested specifying an econometric model in such a way that (thanks to restrictions derived from economic theory) the problem of identification does not arise. The under-determination problem is also strictly connected to the so-called Duhem-Quine thesis: it is not possible to test and falsify a single hypothesis in isolation. This is because any hypothesis is inevitably tied to some auxiliary hypotheses. Auxiliary hypotheses typically include background knowledge, rules of inference, and experimental design that cannot be disentangled from the hypothesis we want to test. Thus, if a particular hypothesis is found to be in conflict with the evidence, we cannot reject the hypothesis with certainty, since we do not know if it is the hypothesis under test or one of the auxiliary hypotheses which is at odds with the evidence. As shown by Sawyer et al. [16], hypothesis testing in economics is further complicated by the approximate nature of theoretical hypotheses. The error in approximation, as well as the less systematic causes disturbing the causal mechanism object of modelling, constitutes an auxiliary hypothesis of typically unknown dimension. For example, in time-series econometric models a distinction is made between 'signal' (which captures the causal mechanisms object of interest) and noise (accounted by the error terms). But it may be the case, as pointed out by Valente [17], that noises are stronger than signals, and that the mechanisms involved undergo several or even continuous structural changes. Econometricians have adopted sophisticated tests which are robust to variations in the auxiliary hypotheses (see, for example, [10]). Nonetheless, the Duhem-Quine thesis still undermines strong apriorist methodologies that do not check the robustness of the empirical results under variations of background assumptions.

18.3 A Taxonomy of the Existing Approaches

A discrete set of approaches for empirical validation, not only different with each other but different to those developed within neoclassical economics, have been developed by the AB community. We suggest that there are two reasons for this heterogeneity. First, AB modellers are interested in phenomena such non-linearities, stochastic dynamics, non-trivial interactions among agents, and feedbacks between the micro and the macro level. These are not amenable to traditional equilibrium modelling approaches and tools. One of the consequences is that AB modellers face an additional set of issues that are not faced by neoclassical modellers. Second, and relatedly, the highly diverse structural content of AB models means they need to be analyzed in very different ways. We propose a taxonomy that maps out the key areas in which AB researchers differ.

The first dimension is the *nature of the objects under study*. This determines the stylised facts (empirically observed facts) that the model is seeking to explain. Significant differences exist with respect to the nature of the objects being studied in AB models. Where neoclassical modellers are interested in quantitative change, AB modellers are equally interested in qualitative change of economic systems themselves. For instance, there are AB models that investigate how R&D spending affects the qualitative nature of macroeconomic growth. Other AB models investigate its quantitative impact, or else seek to explain some statistically observed quantitative property of aggregate growth (e.g. its autocorrelation patterns). Another important distinction is between AB models that seek to investigate a single phenomenon, and those that jointly investigate multiple phenomena. For instance, a model may consider the properties of productivity and investment time-series, in addition to the properties of aggregate growth. Transient versus long-run impact is a further distinction. For example, there are AB models that examine the effect of R&D spending on growth along the diffusion path (the transient) of a newly introduced technology. Other AB models are only concerned with the magnitude of a technologys long-run impact (when the economic system has stabilised somewhat). Finally, an important distinction exists between AB models that investigate micro distributions and macro aggregates. The former are concerned with the dynamics of industry-level distributions, such as a cross-section of firm productivity distributions, for a particular sector, in a particular year. The latter are concerned with longer time-series data for nation states, or the world economy, over a number of years.

A second dimension in which AB models differ is in the *goal of the analysis*. AB models tend to deal with in-sample data. In-sample data is relevant when one is interested in describing or replicating observed phenomena. Out-of-sample exercises, although they are less frequently carried out by AB economists, are essential for the sake of policy evaluation.

A third dimension concerns the nature of the most important *modelling assumptions*. Some models contain many degrees of freedom, others do not. For example, agents in AB models may be characterised by many variables and parameters. Their decision rules may, in turn, be highly-parameterised. Alternatively, agents and decision rules may be described in a very stylised way. Individual decision rule sets and interaction structures may be exogenously fixed. They may change over time. Change may be driven by exogenous, stochastic factors. Alternatively, change may be driven by agents endogenously selecting new decision rules and interaction structures according to some meta-criteria (as it happens in endogenous network formation models, see [6]).

The fourth and final dimension is the *methodology of analysis*. In order to thoroughly assess the properties of an AB model, the researcher needs to perform a detailed sensitivity analysis. This sensitivity analysis should, at the very least, explore how the results depend on (i) micro-macro parameters, (ii) initial conditions, and (iii) across-run variability induced by stochastic elements (e.g. random initial conditions, and random individual decision rules).

There are three important approaches to empirical validation within AB economics: indirect calibration [5], [4], the Werker-Brenner approach to empirical calibration [18], and the history-friendly approach [12], [11].

The *indirect calibration approach* is based on a four-step procedure. In the first step, the modeller identifies a set of stylised facts that (s)he is interested in reproducing and/or explaining with a model. Stylised facts typically concern the macro-level (e.g. the relationship between unemployment rates and GDP growth) but can also relate to cross-sectional regularities (e.g. the shape of the distributions on firm size). In the second step, along with the prescriptions of the empirical calibration procedure, the researcher builds the model in a way that keeps the microeconomic description as close as possible to empirical and experimental evidence about microeconomic behaviour and interactions. This step entails gathering all possible evidence about the underlying principles that inform real-world behaviours (e.g. of firms, consumers, and industries) so that the microeconomic level is modelled in a not-too-unrealistic fashion. In the third step, the empirical evidence on stylised facts is used to restrict the space of parameters, and the initial conditions if the model turns out to be non-ergodic. In the fourth and final step, the researcher should deepen his/her understanding of the causal mechanisms that underlie the stylised facts being studied and/or explore the emergence of 'fresh' stylised facts (i.e. statistical regularities that are different to from the stylised facts of interest), against which the model can be validated ex post). This might be done by further investigating the subspace of parameters that resist to the third step, i.e. those consistent with the stylised facts of interest.

A stream of recent AB contributors to the field of industry and market dynamics has been strongly rooted in the four-step empirical validation procedure just presented. For example, Fagiolo and Dosi [4] study an evolutionary growth model that is able to reproduce many stylised facts about output dynamics, such as I(1) patterns of GNP growth, growth-rates autocorrelation structure, absence of size-effects, etc., while explaining the emergence of self-sustaining growth as the solution of the trade-off between exploitation of existing resources and exploration of new ones. Similarly, Fagiolo et al. [5] present a model of labour and output market dynamics that is not only able to jointly reproduce the Beveridge curve, the Okun curve and the wage curve, but also relates average growth rates of the system to the institutional set-up of the labour market.

The *Werker-Brenner approach* is a three-step procedure for calibrating AB models. The first two steps are consistent with all calibration exercises. The third step is novel. Step 1 uses existing empirical knowledge to calibrate initial conditions and the ranges of model parameters. As mentioned above, AB models contain many dimensions, including the set of assumptions about agents behaviour, their actions, interactions, causal relationships, and the simplifying assumptions of the model. Werker-Brenner propose that, where sensible data are not available, the model should be left as general as possible, i.e. wide ranges should be specified for parameters on which there is little or no reliable data.

Step 2 involves empirical validation of the outputs for each of the model specifications derived from step 1. Through empirical validation, the plausible set of dimensions within the initial dimension space is further reduced. It is possible to run the model specification and generate a Monte Carlo set of micro and macro time-series data for that particular combination of empirically-plausible parameter values. The resulting time-series data — one for each parameter combination — can be thought of as a particular 'theoretical realisation' of the model that is being tested. Of course, any two time-series may overlap to a large extent. This is to be expected since the combinations of parameter values that are being tested are likely to be similar in some dimensions, while different in others. Having generated a set of theoretical realisations for each model specification, one is able to compare these outputs with real-world data. The real-world data that we observe are an 'empirical realisation' that is generated by the rwDGP that we are trying to model. The Werker-Brenner approach advocates the use of Bayesian inference procedures in order to conduct this output validation. Each model specification is assigned a likelihood of being accepted based on the percentage of 'theoretical realisations' that are compatible with each 'empirical realisation.' In this way, empirically observed realisations are used to further restrict the initial set of model specifications (parameter values) that are to be considered. The modeller only retains those parameter values (i.e. model specifications) that are associated to the highest likelihood by the current known facts (i.e. empirical realisations). Model specifications that conflict with current data are discounted.

From a methodological perspective, it is step 3 of the Werker-Brenner approach that is of particular interest. The aim is to find an explanation to the phenomena being studied by exploring the remaining set of model specifications. This is achieved through methodological 'abduction.' Abduction is a process that seeks to describe and explain empirical facts in terms of their underlying structures [18]. In practice, this involves a further validation exercise for all empirical realisations that can be collected. Here, however, the modeller focuses on the shared properties and the characteristics shared by all surviving model specifications in order to identify the invariant properties of the underlying structural model. The authors argue that "these [shared] characteristics can be expected to hold also for the real systems (given the development of the model has not included any crucial and false premises)" [18]. If the characteristics within a group of model specifications differ, then this also offers important insights. "It can be examined which factors in the model are responsible for the differences. Hence, although we will not know the characteristics of the real systems in this case, we will obtain knowledge about which factors cause different characteristics" [18].

While the Weker and Brenner's calibration approach addresses the over-paramete-risation problem by reducing the space of possible 'worlds' that are explored in an AB model, the *history-friendly approach* offers an alternative solution to this problem. Like the calibration approaches discussed above, it seeks to bring modelling more closely 'in line with the empirical evidence' and thereby constrains the analysis to reduce the dimensionality of a model. The key difference is that this approach uses the specific historical case studies of an industry to model parameters, agent interactions, and agent decision rules. In effect, it is a calibration approach which uses particular historical traces in order to calibrate a model.

In part, the history-friendly approach represents an attempt to deal with criticisms levelled at early neo-Schumpeterian AB models of technological change. Two of the key protagonists of history-friendly modelling, R. Nelson and S. Winter, were founding fathers of neo-Schumpeterian AB modelling. While the early models were much more micro-founded and empirically-driven than contemporary neoclassical models, empirical validation was weak. There was a lack of thorough sensitivity and validation checks and empirical validation, when carried out, tended to consist of little more than a cursory comparison of outputs generated by a just a handful of simulation runs with some very general stylised facts. Further, the early models contained many dimensions and so it was rather easy to generate a few outputs that matched some very general observations (the over-parameterisation problem).

In terms of our taxonomy, the history-friendly approach is strongly quantitative and mainly focuses on microeconomic transients (industrial paths of development). In this approach a good model is one that can generate multiple stylised facts observed in an industry. The approach has been developed in a series of papers. Key amongst these are [12] and [11]. In [12], Malerba, Nelson, Orsenigo and Winter outlined the approach and then applied it to a discussion of the transition in the computer industry from mainframes to desktop PCs. In [11], the approach was applied to the pharmaceutical industry and the role of biotech firms therein. Through the construction of industry-based AB models, detailed empirical data on an industry informs the AB researcher in model building, analysis and validation. Models are to be built upon a range of available data, from detailed empirical studies to anecdotal evidence

Table 18.1. Taxonomy of dimensions of heterogeneity in empirical validations of AB models

Approach	Domain of Application	Which kind of data should one employ?	How to employ data?	What to do first?
Indirect Calibration	-Micro (industries, markets) -Macro (countries, world economy)	-Empirical data	-Assisting in model building -Validating simulated output	-First validate, then indirectly calibrate
Werker-Brenner	-Micro (industries, markets) -Macro (countries, world economy)	-Empirical data -Historical knowledge	-Assisting in model building -Calibrate initial conditions and parameters -Validating simulated output	-First calibrate, then validate
History-Friendly	-Micro (industries, markets)	-Empirical data -Casual, historical and anecdotic knowledge	-Assisting in model building -Calibrate initial conditions and parameters -Validating simulated output	-First calibrate, then validate

to histories written about the industry under study. This range of data is used to assist model building and validation. It should guide the specification of agents (their behaviour, decision rules, and interactions), and the environment in which they operate. The data should also assist the identification of initial conditions and parameters on key variables likely to generate the observed history. Finally, the data are to be used to empirically validate the model by comparing its output (the simulated trace history) with the actual history of the industry. It is the latter that truly distinguishes the history-friendly approach from other approaches. Previous researchers have used historical case studies to guide the specification of agents and environment, and to identify possible key parameters. The authors of the history-friendly approach suggest that, through a process of backward induction one can arrive at the correct set of structural assumptions, parameter settings, and initial conditions. Having identified the correct set of 'history-replicating parameters,' one can carry on and conduct sensitivity analysis to establish whether (in the authors' words) 'history divergent' results are possible.

Table 18.1 summarizes the main characteristics of the three different approaches. The first dimension, in which these approaches differ, is the domain of application. The direct and indirect calibration approaches can, in principle, be applied to micro and macro AB models (e.g. to describe the dynamics of firms, industries, and countries). By contrast, the history-friendly approach only addresses micro dynamics. A second dimension of heterogeneity is the type of data that are used for empirical validation. In addition to empirical datasets, the Werker-Brenner approach advocates the use of historical knowledge. The history-friendly approach allows one to employ casual and anecdotic knowledge as well. The third dimension is the way in which data is actually used. All three approaches use data to assist model building, as well as val-

idating the validation of the simulated outputs of models. Unlike the other
two approaches, indirect calibration does not directly employ data to calibrate initial conditions and parameters. The fourth dimension is the order in
which validation and calibration is performed. Both the Werker-Brenner and
the history-friendly approaches first perform calibration and then validation.
By contrast, the indirect calibration approach first performs validation, and
then indirectly calibrates the model by focusing on the parameters that are
consistent with output validation.

18.4 Open-ended Issues and Conclusions

There is a set of core issues that affect all the approaches and which (so far)
remain unresolved. In this concluding section we shed some light on that.

1. *Alternative strategies for constructing empirically-based models.* There
is intense debate about the best way to actually construct empirically-based
models, and to select between alternative models. What happens, for instance,
if there are alternative assumptions and existing empirical data does not assist
in choosing between them? A number of different strategies exist for selecting assumptions in the early stages of model building [3]. One strategy is
to start with the simplest possible model, and then proceed to complicate
the model step-by-step. This is the KISS strategy: 'Keep it simple, stupid!' A
very different strategy is the KIDS strategy: 'Keep it descriptive, stupid!' Here
one begins with the most descriptive model one can imagine, and then simplify it as much as possible. The third strategy, common amongst neoclassical
economists, is TAPAS: 'Take A Previous model and Add Something.' Here
one takes an existing model and successively explores the assumption space
through incremental additions and/or the relaxation of initial assumptions.

2. *Problems that arise as a consequence of over-parameterisation in AB
models.* Whatever the strategy employed, the AB modeller often faces an
over-parameter-isation problem. AB models with realistic assumptions and
agent descriptions invariably contain many degrees of freedom. There are two
aspects to the over-parameterisation problem. Firstly, the dimensions of the
model may be so numerous that it can generate any result. If this is the case,
then the explanative potential of the model is little better than a random
walk. Secondly, the causal relations between assumptions and results become
increasingly difficult to study. A possible strategy is to use empirical evidence
to restrict the degrees of freedom, by directly calibrating initial conditions
and/or parameters. Then, one can indirectly calibrate the model by focussing
on the subspace of parameters and initial conditions under which the model
is able to replicate a set of stylised facts. Unfortunately, this procedure still
tends to leave the modeller with multiple possible 'worlds.'

3. *The usefulness and implications of counterfactuals for policy analysis*
How does one interpret the counterfactual outputs generated by a model? It

is tempting to suggest that outputs which do not accord with empirical observations are counterfactuals, and that the study of these counterfactuals are useful for policy analysis. Cowan and Foray [2] suggest that it is exceedingly difficult, in practice, to construct counterfactual histories because economic systems are stochastic, non-ergodic, and structurally evolve over time. As AB models typically include all these elements in their structure, Cowan and Foray argue that using (evolutionary) AB models to address counterfactual-like questions may well be misleading. More generally, comparing the outputs generated by AB models with real-world observations involves a set of very intricate issues. For example, Windrum [19] observes that the uniqueness of historical events sets up a whole series of problems. In order to move beyond the study of individual traces, we need to know if the distribution of output traces generated by the model mDGP approximates the actual historical traces generated by the rwDGP under investigation. A way to circumvent the uniqueness problem is to employ a strong invariance assumption on the rwDGP, thereby pooling data that should otherwise be considered a set of unique observations. For example, one typically supposes that cross-country aggregate output growth rates come from the same DGP. Similarly, it is supposed that the process that driving firm growth does not change across industries or time (up to some mean or variance scaling). This allows one to build cross-section and time-series panel data. Unfortunately we cannot know if the suppositions are valid. But this is often not possible in practice. Consider the following example. Suppose the rwDGP in a particular industry does not change over time (i.e. it is ergodic). Even if this is the case, we do not typically observe the entire distribution of all observations but rather a very limited set of observations — possibly only one, unique roll of the dice. The actual history of the industry we observe is only one of a set of possible worlds. So how do we know that the actual historical trace is in any sense typical (statistically speaking) of the potential distribution? If we do not know this, then we have nothing against which to compare the distributions generated by our model. We cannot determine what is typical, and what is atypical.

4. *Definition of sufficiently strong empirical tests.* The fundamental difficulties in defining strong tests for model outputs is highlighted by Brock's [1] discussion of 'unconditional objects' in economics. Empirical regularities need to be handled with care because we only have information on the properties of stationary distributions. The data that we observe does not provide information on the dynamics of the stochastic processes that actually generated them. Therefore, replication does not necessary imply explanation. For example, many evolutionary growth models can generate similar outputs on differential growth-rates between countries, technology leadership and catch-up, even though they differ significantly with respect to the behaviour and learning procedures of agents, and in their causal mechanisms [19]. Similarly, the Nelson and Winter [15] model replicates highly aggregated data on time paths for output (GDP), capital and labour inputs, and wages (labour share in output), but these outputs can also be replicated by conventional neoclassi-

cal growth models. In the same vein, there might be many different stochastic processes (and therefore industry dynamic models) that are able to generate, as a stationary state, a power-law distribution for the cross-section firm size distribution. Although one may be unable to narrow down a single model, we may be able to learn about the general forces at work, and to restrict the number of models that can generate a set of statistical regularities [1]. Therefore, as long as the set of stylised facts to be jointly replicated is sufficiently large, any 'indirect' validation could be sufficiently informative, because it can effectively help in restricting the set of all stochastic processes that could have generated the data displaying those stylised facts. Another way out the conditional objects critique would be to not only validate the macro-economic output of the model, but also its micro-economics structure, e.g. agents behavioural rules. This requires one to only include in the model individual decision rules (e.g. learning) that have been validated by empirical evidence. Of course, this would require highly detailed and reliable data about micro-economic variables, possibly derived from extensive laboratory experiments.

5. *Availability, quality and bias of datasets.* Empirically-based modelling depends on high quality datasets. Unfortunately, the datasets that exist are invariably pre-selected. Not all potential records are retained; some are fortuitously bequeathed by the past but others are not captured. Datasets are constructed according to criteria that reflect certain choices and, as a consequence, are biased. As econometricians know only too well, it may simply be the case that data that would have assisted in a particular discussion has simply not been collected. A further and often neglected problem is that standard econometric methods are influenced by prevailing theoretical orthodoxy.

References

[1] Brock W (1999) Scaling in economics: a reader's guide, Industrial and Corporate Change, 8: 409-446

[2] Cowan R., Foray D (2002) Evolutionary economics and the counterfactual threat: on the nature and role of counterfactual history as an empirical tool in economics, Journal of Evolutionary Economics, 1 2: 539-562

[3] Edmonds B, Moss S (2005) From KISS to KIDS - an 'anti-simplistic' modelling approach. In Davidsson P, Logan B, Takadama K (eds.): Multi Agent Based Simulation 2004. Springer, Lecture Notes in Artificial Intelligence, 3415:130-144

[4] Fagiolo G, Dosi G (2003) Exploitation, Exploration and Innovation in a Model of Endogenous Growth with Locally Interacting Agents, Structural Change and Economic Dynamics, 14: 237-273

[5] Fagiolo G, Dosi G, Gabriele R (2004a) Matching, Bargaining, and Wage Setting in an Evolutionary Model of Labor Market and Output Dynamics, Advances in Complex Systems, 14: 237-273

[6] Fagiolo G, Marengo L, Valente M (2004b) Endogenous Networks in Random Population Games, Mathematical Population Studies, 11: 121-147

[7] Friedman M (1953) The Methodology of Positive Economics, in Essays in Positive Economics, Chicago: University of Chicago Press

[8] Haavelmo T (1944) The Probability Approach in Econometrics, Econometrica, 12: 1-115

[9] Lakatos I (1970) Falsification and the Methodology of Scientific Research Programmes, in I Lakatos and A. Musgrave, Criticism and the Growth of Knowledge, Cambridge: Cambridge University Press, pp. 91-196

[10] Leamer EE (1978) Specification Searches, Ad Hoc Inference with Non-experimental Data, New York: John Wiley

[11] Malerba F, Orsenigo L (2001) Innovation and market structure in the dynamics of the pharmaceutical industry and biotechnology: towards a history friendly model, Conference in Honour of Richard Nelson and Sydney Winter, Aalborg, 12th - 15th June 2001

[12] Malerba F, Nelson RR, Orsenigo L, Winter SG (1999) History friendly models of industry evolution: the computer industry, Industrial and Corporate Change, 8: 3-41

[13] Mäki U (1992) On the Method of Isolation in Economics, Poznan Studies in the Philosophy of the Sciences and the Humanities, 26:19-54

[14] Mäki U (1998) Realism, in JB Davis, D Wade Hands U Mäki (eds.), The Handbook of Economic Methodology, Cheltenham, UK: Edward Elgar, pp. 404-409

[15] Nelson R, Winter S (1982) An Evolutionary Theory of Economic Change, Harvard University Press, Cambridge

[16] Sawyer K R, Beed C, Sankey H (1997) Underdetermination in Economics. The Duhem-Quine Thesis, Economics and Philosophy, 13: 1-23

[17] Valente M (2005) Qualitative Simulation Modelling, Faculty of Economics, University of L'Aquila, L'Aquila, Italy, mimeo

[18] Werker C, Brenner T (2004) Empirical Calibration of Simulated Models, Papers on Economics and Evolution 0410, Max Planck Institute of Economics, Jena

[19] Windrum P (2004) Neo-Schumpeterian simulation models, forthcoming in H Hanusch, A Pyka (eds.) The Elgar Companion to Neo-Schumpeterian Economics, Edward Elgar: Cheltenham

Equilibrium Return and Agents' Survival in a Multiperiod Asset Market: Analytic Support of a Simulation Model

Mikhail Anufriev[1] and Pietro Dindo[2]

[1] CeNDEF, University of Amsterdam, Amsterdam m.anufriev@uva.nl
[2] CeNDEF, University of Amsterdam, Amsterdam p.d.e.dindo@uva.nl

Summary. We provide explanations for the results of the Levy, Levy and Solomon model, a recent simulation model of financial markets. These explanations are based upon mathematical analysis of a dynamic model of a market with an arbitrary number of heterogeneous investors allocating their wealth between two assets. The investors' choices are endogenously modeled in a general way and, in particular, consistent with the maximization of an expected utility. We characterize the equilibria of the model and their stability and discuss implications for the market return and agents' survival. These implications are in agreement with the results of previous simulations. Thus, our analytic approach allows to explore the robustness of the previous analysis and to expand its spectrum.

19.1 Introduction

The goal of this paper is to explore analytically the framework underlying simulations of the so-called Levy, Levy and Solomon (henceforth LLS) model. The model was introduced in [6] and further results were presented in [7] and [9], among others. See also [8] for extensive discussion. The motivation behind the model was to investigate whether some financial anomalies (like excess volatility or autocorrelation of returns) can be explained by relaxing the traditional assumption of classical finance about the presence of a fully-informed and rational representative agent. The LLS framework assumes the presence of heterogeneous agents whose market impact depends on their past performances. In the words of its authors ([8], p. 143):

> "The LLS model incorporates some of the main empirical findings regarding investor behavior, and we employ this model to study the effect of each element of investor behavior on asset pricing and market dynamics."

The model has been shown to qualitatively explain many of the financial anomalies, but all its results are based on simulations. The criticism of the simulation approach usually points at a huge number of degrees of freedom, i.e. dimensions of a set consisting of (i) all possible parameters, (ii) realizations of the random variables and (iii) initial conditions. This leads to the feeling that "everything one wants to obtain" can be obtained in the heterogeneous world. In other words, the absence of a closed form solution makes it difficult to believe that the results are robust. As a reply to that criticism, many analytic models of financial markets with heterogeneous agents appeared, see [5] for a recent review. On the other hand, an analytic approach is limited due to the high non-linearity of the models with heterogeneous agents. For example, the agents' wealth evolution is usually neglected in analytic contributions. Therefore, both analytic and simulation approaches have to co-exist and to supplement each other. As we show with our analysis, the analytic investigations of the LLS model can effectively supplement the results of previous simulation exercises.

Our analytic model of the LLS framework starts off with a pure exchange, two-assets economy, where agents invest according to different rules. The framework is consistent with the CRRA (Constant Relative Risk Aversion) behavior, so that the individual demand for the risky asset is expressed as a fraction of the agent's wealth. Consequently, the price and agents' wealths are determined simultaneously, and, moreover, agents with different wealth levels have different impact on the price realization.

Models in [2, 3, 4] are predecessors of our model. In particular, as in [3], we model the agents' behavior by means of generic *investment functions*, mapping the available information on the current investment choice. However, we substantially deviate from these papers since we introduce a more realistic dividend process. Instead of assuming a constant dividend yield, we analyse the case where the dividend is growing at a given constant rate. This system corresponds to the *deterministic skeleton* of a market where dividend follows a geometric random walk. We provide equilibrium and stability analysis for this skeleton, which sheds light on the behavior of the stochastic LLS model, where the growth rate of dividends is random.

The direct application of our analytic model follows from the fact that the market structure we use is the same as in simulations of the LLS model. In [6, 7, 8, 9] the agents are expected utility maximizers having power utility function. One of the obstacles on the way to explore such setting analytically is the absence of a closed-form solution for the corresponding optimization problem. This obstacle has played a role in arguments in favor of simulations. However, in our framework with investment functions the precise solutions are not necessary, since the analytic results are expressed in terms of the general functions and can be illustrated geometrically. The difficulty of dealing with a power utility function is overcome, and comparative statics exercises can be easily performed, analogously to what has been done in [1]. Thus, our analysis allows to explain simulation results that alternatively have to be described in an rather vague fashion as in the following quote from [9] (p.568, 569):

"Looking more systematically at the interplay of risk aversion and memory span, it seems to us that the former is the more relevant factor, as with different [risk aversion coefficients] we frequently found a reversal in the dominance pattern: groups which were fading away before became dominant when we reduced their degree of risk aversion. [...] It also appears that when adding different degrees of risk aversion, the differences of time horizons are not decisive any more, provided the time horizon is not too short."

The rest of the paper is organized as follows. In Section 19.2 the analytic model is presented. In Section 19.3 the main results of the equilibrium and stability analysis are summarized in a few propositions. In Section 19.4 we apply these results and, therefore, offer a rigorous explanation of the findings in [7] and [9], among others. The analytic results also help to discuss the robustness of the simulation results with respect to the different assumptions. We also present some further results in order to characterize the dynamics when the equilibria are unstable. Section 19.5 concludes.

19.2 Model Structure

Let us consider N agents trading in discrete time in a two-asset economy with a riskless asset giving a constant interest rate $r_f > 0$ and constant supply (normalized to 1) of risky asset paying a random dividend D_t. The price of the riskless asset is fixed to 1, and the price P_t of the risky asset is fixed through market clearing. Let $W_{t,n}$ stand for the wealth of agent n at time t and $x_{t,n}$ for its share invested in the risky asset. The dividend is paid before trade starts, so the wealth evolves as

$$W_{t+1,n} = (1 - x_{t,n}) W_{t,n} (1 + r_f) + \frac{x_{t,n} W_{t,n}}{P_t} (P_{t+1} + D_{t+1}). \quad (19.1)$$

The price at time t is fixed through the market clearing condition

$$\sum_{n=1}^{N} x_{t,n} W_{t,n} = P_t. \quad (19.2)$$

Assume that the agent's investment share $x_{t,n}$ does not depend upon the wealth. The resulting demand is consistent with the one derived from the maximization of a constant relative risk aversion (CRRA) utility function. Moreover the investment shares are independent of the contemporaneous price and bounded between zero and one, $x_{t,n} \in (0,1)$, for all t and n. Both assumptions are consistent with previous simulations of the LLS model and simplify the analysis substantially. Notice that according to (19.1), the wealth does depend upon the contemporaneous price, so that price and wealth are simultaneously determined by the market clearing condition (19.2). Thus, (19.1) and (19.2) give the evolution of the state variables $W_{t,n}$ and P_t over time implicitly, provided that the investment shares $\{x_{t,n}\}$ are specified.

Concerning the latter we further assume that for each agent n there exists an *investment function* f_n such that

$$x_{t,n} = f_n(\mathcal{I}_t), \qquad (19.3)$$

where $\mathcal{I}_t = \{D_t, D_{t-1}, \ldots, P_{t-1}, P_{t-2}, \ldots\}$ is the information set available to the agents at time t. Agents' investment decisions evolve following individual prescriptions. The generality of the investment functions allows a big flexibility in the modeling of the agents' behaviors. Formulation (19.3) includes as special cases both technical trading, when agents' decisions are driven by the observed price fluctuations, and more fundamental attitudes, e.g. when the decisions are made on the basis of the price-dividend ratio. It also includes the case of constant investment strategy, occurring when agent assumes the stationarity of the *ex-ante* return distribution.

For our application in Section 19.4 it is important to stress that (19.3) includes those investment behaviors which are derived from expected utility maximization with power utility $U(W, \gamma) = W^{1-\gamma}/(1 - \gamma)$, where $\gamma > 0$ is the relative risk aversion coefficient. Indeed, solution of such a problem has a wealth independent investment share. This property holds for any distribution of the next period return which the agent is assumed to perceive now and for any risk aversion. However, the solution is unavailable in explicit form. Consequently, the analysis of the LLS model in [6, 7, 8, 9] rely on numeric solutions. Since the results of Section 19.3 are valid for any given functional form f, provided some easy-to-check general properties, we are able to perform an analytic analysis of the LLS model even when agents maximize expected utility with power utility function.

Accordingly with the LLS model, assume that $D_t = D_{t-1}(1 + \tilde{g})$, where the growth rate, $\tilde{g}$, is an i.i.d. random variable whose mean is g. Below we perform an analysis of the *deterministic skeleton* of the dynamics triggered by this stochastic process, and we fix the growth rate of dividends to a constant value g.

With some algebra one can show that the implicit dynamics described in (19.1) and (19.2) can be made explicit. The resulting system is written in terms of the price return $k_{t+1} = P_{t+1}/P_t - 1$, dividend yield $y_{t+1} = D_{t+1}/P_t$ and agents' relative wealth shares in the aggregate wealth $\varphi_{t,n} = W_{t,n}/\sum_m W_{t,m}$ as follows

$$
\begin{cases}
y_{t+1} = y_t \dfrac{1 + g}{1 + k_t} \\[2ex]
k_{t+1} = r_f + \dfrac{\sum_m \big((1 + r_f)(x_{t+1,m} - x_{t,m}) + y_{t+1}\, x_{t,m}\, x_{t+1,m}\big)\varphi_{t,m}}{\sum_m x_{t,m}(1 - x_{t+1,m})\,\varphi_{t,m}} \\[2ex]
\varphi_{t+1,n} = \varphi_{t,n} \dfrac{(1 + r_f) + (k_{t+1} + y_{t+1} - r_f)\, x_{t,n}}{(1 + r_f) + (k_{t+1} + y_{t+1} - r_f)\sum_m x_{t,m}\varphi_{t,m}} \\[2ex]
x_{t+1,n} = f_n\big(k_t, k_{t-1}, \ldots, k_{t-L}; y_{t+1}, y_t, \ldots, y_{t-L}\big)
\end{cases} \cdot
$$

$$\qquad (19.4)$$

The numerator of the fraction in the right-hand side of the third equation in (19.4) represents the wealth return of agent n. Thus, the relative wealth changes in accordance with the agent's performance relative to the average performance, where the return of individual wealth should be taken as a performance measure. The second equation in (19.4) stresses the role of the agents' relative wealths in the return determination: the richer agents have higher influence on the market. Finally, the last equation in (19.4) specifies the information set $\mathcal{I}_t$ in terms of the same variables as other equations. For the further analysis we assume that agents base their behavior on the finite number of past price returns and dividend yields. Their memory span L can be arbitrarily large, however.

19.3 Equilibrium Return and Agents' Survival

Given the arbitrariness of the size of population N and absence of any specification for the investment function, the analysis of the dynamic behavior generated by system (19.4) is highly non-trivial in its general formulation. One may, indeed, expect that nothing specific can be said about the dynamics. Let us, however, limit ourselves to the "equilibrium" situations, corresponding to the fixed points of system (19.4). In this Section we investigate how such equilibria can be characterized, under which conditions they represent the long-run behavior of the system (in other words, when they are stable), and which agents have positive wealth shares, i.e. survive, in the equilibria. The proofs of all statements are available upon request.

19.3.1 Location of Equilibria

The following result allows us to classify all possible equilibria into two classes, depending upon the values of two exogenous variables.

Proposition 1. *Let us consider the equilibrium of the system (19.4) given by the dividend yield y^*, return k^*, investment shares $(x_1^*, \ldots, x_n^*)$ and wealth distribution $(\varphi_1^*, \ldots, \varphi_n^*)$.*
The two following cases are possible:

(i). $g > r_f$. Then $k^* = g$, and all survivors (agents with non-zero wealth shares) have the same investment share $x_\diamond^*$, which together with y^* satisfies

$$\frac{g - r_f}{y^*} = \frac{x_\diamond^*}{1 - x_\diamond^*}. \qquad (19.5)$$

(ii). $g \leq r_f$. Then $k^* = r_f$ and $y^* = 0$.

In both cases the wealth shares of survivors are arbitrary positive numbers summing to 1, while the agent's investment shares satisfy $x_n^* = f_n(k^*, \ldots, k^*; y^*, \ldots, y^*)$, with corresponding k^* and y^*.

This result shows that the equilibrium price return is $k^* = \max(g, r_f)$. If the dividend growth rate is smaller than r_f, the dividend yield converges to zero, and the risky asset asymptotically yields the same return as the riskless asset. In this case, the equilibria described in Proposition 1 *(ii)* are referred as **no-equity premium equilibria** (NEPE). The investment shares of agents are unambiguously determined through the investment functions, while the wealth shares are free of choice, so any number of agents can survive in such equilibria. Notice that NEPE imply zero dividend yield and, therefore, are unfeasible, strictly speaking. They can be observed asymptotically, however.

If the dividend grows fast enough, so that $g > r_f$, the equilibrium dividend yield y^* depends on agents' behaviors. From (19.5) one can easily show that the risk premium in such an equilibrium is positive and equal to $(g - r_f)/x_\diamond^*$. Consequently, the equilibria from Proposition 1 *(i)* are called the **equity premium equilibria** (EPE). Even if the EPE can have any number $M \in \{1, \ldots, N\}$ of survivors, all of them must behave identically and invest $x_\diamond^*$. This is the key result for getting a simple geometric characterization of the EPE. Indeed, it implies that all possible couples "dividend yield – survivor's investment share" belong to a one-dimensional curve, which is introduced below.

Definition 1. *The Equilibrium Market Line (EML) is the following function*

$$l(y) = \frac{g - r_f}{y + g - r_f} \qquad defined\ for \quad y > 0. \tag{19.6}$$

Now it follows from (19.5), that the dividend yield in the EPE with M survivors (which are the first M agents, without loss of generality) should satisfy to M equations

$$l(y^*) = f_n(g, \ldots, g; y^*, \ldots, y^*) \qquad \forall n \in \{1, \ldots, M\}.$$

In other words, the dividend yield in the EPE can be found as an intersection of the EML with M one-dimensional functions representing the "diagonal" cross-sections of the original investment functions by the set

$$\left\{ k_t = k_{t-1} = \cdots = k_{t-L} = g; \quad y_{t+1} = y_t = \cdots = y_{t-L} = y \right\}. \tag{19.7}$$

The left panel of Fig. 19.1 illustrates the EPE in the market with two different agents, whose investment functions (more precisely, diagonal cross-sections of the original investment functions) are shown as thin lines marked as I and II. Their three intersections with the EML, shown as a thick line, give all the possible EPE. At equilibrium S the agent I is the only survivor, so that $\varphi_1^* = 1$. The dividend yield y^* at this equilibrium is the abscissa of the point S, while the investment share of the survivor, x_1^*, is the ordinate of S. Finally, the investment share of the second agent can be found as a value of his investment function at y^*. Notice that in this equilibrium $x_1^* > x_2^*$. Analogously, the variables are determined in other two equilibria. In particular, agent I is the

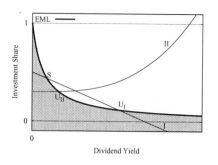 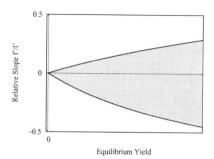

Fig. 19.1. Location and stability of equilibria for $g > r_f$. **Left Panel:** EPE are intersections of the EML with the investment functions; **Right Panel:** Stability of equilibria for $L = 1$.

only survivor at equilibrium U_I, while at U_{II} the second agent survives, $\varphi_2^* = 1$.

In all equilibria illustrated in Fig. 19.1 only one agent survives. In the case of more then one survivors, Proposition 1(i) implies that their investment functions should have a common intersection with the EML. Such situation is rather special, while the illustrated example can be classified as "generic".

Finally, with some simple algebra, one can characterize the agents' wealth growth rates in different equilibria.

Corollary 1. (i). *The wealth return of agent n is equal to $1 + r_f + x_n^*(g - r_f)/x_\diamond^*$ at any EPE. Thus, the wealth of all survivors grows at the same rate g.*
(ii). *At the NEPE the wealth of all the agents grows at the same rate r_f.*

19.3.2 Stability of Equilibria

The next natural question concerns the stability of the equilibria characterized in Proposition 1. In this paper we investigate this question only for the case $g > r_f$, i.e. only for the equity premium equilibria. The following general result holds.

Proposition 2. *The EPE, described in Proposition 1(i), where the first M agents survive, is stable if and only if the following conditions are met:*
1) the equilibrium investment shares of the non-surviving agents are such that

$$x_\diamond^* \left(1 - 2(1 + g)/(g - r_f)\right) < x_m^* < x_\diamond^* \qquad \forall m \in \{M + 1, \ldots, N\}. \quad (19.8)$$

2) after eliminating all non-surviving agents, the behavior of survivors generates stable dynamics.

This Proposition gives an important necessary condition for stability of the EPE. Namely, investment shares of non-surviving agents must satisfy (19.8). The leftmost inequality is always satisfied for reasonable values of g and r_f, while the rightmost inequality shows that the survivors should behave more aggressively in equilibrium, i.e. invest higher investment share, than those who do not survive. This result is intuitively clear, because, according to Corollary 1, the most aggressive agent has a higher wealth return at the EPE. Proposition 2 implies the instability of equilibria U_I and U_{II} in the example shown on the left panel of Fig. 19.1. In the stable equilibrium the investment shares of non-surviving agents should belong to the gray area.

If condition (19.8) is satisfied, the non-survivors can be eliminated from the market. When is the resulting equilibrium stable? We answer this question only for the case of single survivor with investment function dependent upon the average of past L total returns

$$x_t = f\left(\sum_{\tau=1}^{L} (y_{t-\tau} + k_{t-\tau})/L\right). \tag{19.9}$$

This special case will be important in the applications of Section 19.4. Standard stability analysis leads to the following result.

Proposition 3. *Let (x^*, y^*, k^*) be an EPE with one survival agent. The EPE is asymptotically stable if and only if all the roots of polynomial*

$$Q(\mu) = \mu^{L+1} - \frac{1 + \mu + \cdots + \mu^{L-1}}{L}\left(1 + (1-\mu)\frac{1+g}{y}\right)\frac{f'(y^*+g)}{l'(y^*)} \tag{19.10}$$

lie inside the unit circle.

From Section 19.3.1 it follows that the equilibrium yield at the EPE is given as a solution of $l(y) = f(y + g)$. Thus, the last fraction in the polynomial (19.10) gives the relative slope of the investment function and the EML at the equilibrium. On the EML plot, this is the relative slope of the cross-section of an investment function and the EML in the intersection.

Propositions 2 and 3, give exhaustive characteristics of stability conditions of the EPE with single survivor in the market where agents behave according to (19.9). The stability conditions are implicit, however, since they contain a requirement on the roots of polynomial $Q(\mu)$. When $L = 1$ this requirement can be made explicit. Namely, the following two inequalities are sufficient for stability:

$$\frac{f'(y^*+g)}{l'(y^*)} > \frac{-y^*}{1+g+y^*} \quad \text{and} \quad \frac{f'(y^*+g)}{l'(y^*)} < \frac{y^*}{y^* + 2(1+g)}.$$

These conditions are illustrated in the right panel of Fig. 19.1 in the coordinates $(y^*, f'/l')$. The equilibrium is stable if it belongs to the gray area.

A mixture of analytic and numeric tools helps to reveal the behavior of the roots of polynomial (19.10) with higher L, and, therefore, to understand the impact of the agent's memory span on the stability of corresponding equilibrium. In general, the equilibrium stabilizes with lower (in absolute value) relative slope f'/l' at the equilibrium and with higher memory span L.

19.4 Analytic Support of Simulations

All the simulations of the LLS model deal with agents who maximize a power utility function with relative risk aversion γ, and who use the average of the last L returns as an estimate for the next period return. Even if the investment function for such an optimization problem cannot be derived explicitly, one can investigate how the cross-section of this function by the hyperplane (19.7) changes with parameters γ and L. In this Section we show that this is sufficient for explaining the results of the simulations in [6, 7, 8, 9]. We start the analysis by illustrating the effects of the risk aversion and memory span in the case of mean-variance investment function (which can be derived explicitly). The insights developed in this case will then be used to discuss the results of the original simulations. Throughout this Section it is assumed that $g > r_f$, so that only EPE are analyzed.

Let us consider an agent who maximizes the following mean-variance utility

$$U = E_t[x_t(k_{t+1} + y_{t+1}) + (1 - x_t)r_f] - \frac{\gamma}{2}V_t[x_t(k_{t+1} + y_{t+1})], \qquad (19.11)$$

where E_t and V_t denote, respectively, the mean and the variance conditional on the information available at time t, and γ is the coefficient of risk aversion. Assuming constant expected variance $V_t = \sigma^2$, the optimal investment fraction is $x_t^* = E_t[k_{t+1} + y_{t+1} - r_f]/(\gamma\sigma^2)$. Consistently with the LLS framework, we assume that the next period return is estimated as the average of past L realized return, while the expected variance is constant, and we bound the investment shares in the interval $[0.01, 0.99]$. Thus, the investment function reads

$$f_{\alpha,L} = \min\left\{0.99, \max\left\{0.01, \frac{1}{\alpha}\left(\frac{1}{L}\sum_{\tau=1}^{L}(k_{t-\tau} + y_{t-\tau}) - r_f\right)\right\}\right\},$$

where we have defined $\alpha = \gamma\sigma^2$. From Section 19.3.1 it follows that all the EPE can be found as the intersections of the EML with the function

$$\tilde{f}_\alpha(y) = \min\left\{0.99, \max\left\{0.01, \frac{y + g - r_f}{\alpha}\right\}\right\},$$

which is the cross-section of $f_{\alpha,L}$ by the hyperplane (19.7). The left panel of Fig. 19.2 illustrates the situation for a single agent. The market has a unique equilibrium, A_α, whose abscissa, y_α^*, is the equilibrium dividend yield, and whose ordinate, x_α^*, is the equilibrium agent's investment share. This

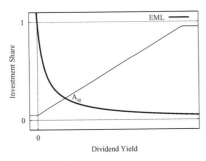 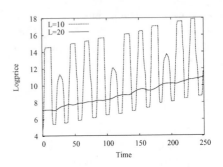

Fig. 19.2. Market dynamics with a single mean-variance maximizer. **Left panel:** The equilibrium on the EML. **Right panel:** Log-price dynamics over the simulations for two different values of memory span L. In both cases the investment function is the one depicted in the left panel.

equilibrium does not depend on the memory span L, but depends on the (normalized) risk aversion coefficient α. When α increases, the line $x = (y + g - r_f)/\alpha$ rotates counter-clock wise. Therefore, the equilibrium yield is an increasing function of the risk aversion, while the equilibrium investment share is a decreasing function of the risk aversion.

What are the determinants of the stability of the equilibrium A_α? First of all, notice that the stability analysis of Section 19.3.2 can be applied, because the investment function $f_{\alpha,L}$ is of the type specified in (19.9). The stability, therefore, is determined both by the relative slope of the function $\tilde{f}_\alpha$ with respect to the EML in point A_α and by the memory span L. In particular, the increase of L brings stability to the system.

The right panel of Fig. 19.2 shows the log-price time series resulting from two simulations of the model for the investment function $\tilde{f}_\alpha$ in the case where the dividend follows a geometric random walk. The only difference between simulations lies in the memory span L. The dotted line shows dynamics for the agent with $L = 10$. The equilibrium is unstable in this case, and the endogenous fluctuations which we observe are determined by the upper and lower bounds of $\tilde{f}_\alpha$. Moreover, the period of fluctuations is related to L. The solid line shows the price series obtained with memory increased to $L = 20$. The system converges to the stable equilibrium, and the fluctuations are due to exogenous noise affecting the dividend growth rate. Notice that a different α value may require a different minimum value of L to produce a stable equilibrium.

We now turn to the analysis of a market with many agents. In this case we are particularly interested in assessing the agents' survival. For this purpose we use the results of Proposition 2, namely that the survivor should have the highest investment share at his intersection with the EML. If, being alone, the survivor generates a stable equilibrium, he also dominates the market, i.e. asymptotically has all the wealth. The top left panel of Fig. 19.3 shows

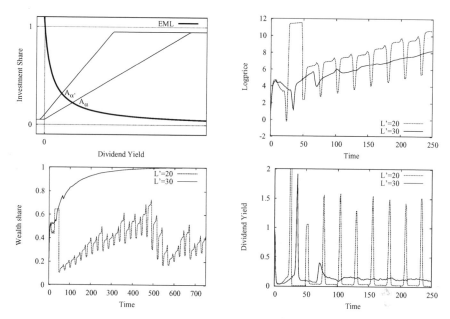

Fig. 19.3. Market dynamics with two mean-variance maximizers. **Top left panel:** EML and two investment functions. The agent with lower risk aversion α' produces equilibrium A'. **Top right panel:** Log-price dynamics. **Bottom left panel:** Dynamics of the wealth share of the agent with lower risk aversion α'. **Bottom right panel:** The dividend yield dynamics. All simulations are performed for two levels of memory span of the agent with low risk aversion.

two investment functions for different values of risk aversion, namely α and $\alpha' < \alpha$. Since at y_α^* the agent with risk aversion α invests less then the other agent, he cannot survive at "his" equilibrium, A_α, and, therefore, he can never dominate the market. Whether the agent with risk aversion α' is able to dominate the market depends on the stability of "his" equilibrium, $A_{\alpha'}$. If the memory he uses is long enough, the equilibrium is stable and $\varphi_{\alpha'}^* = 1$.

Fig. 19.3 shows the results of simulations for two different values of the memory parameter L' for the agent with lower risk aversion α'. When $L' = 20$, this agent, while destroying the previously stable equilibrium A_α does not bring the system to a new equilibrium. In fact, he destabilizes the price dynamics but fails to dominate the market and his wealth share keeps fluctuating between zero and one. However, when the memory of this agent increases to $L' = 30$, the new equilibrium $A_{\alpha'}$ is stabilized and he ultimately dominates the market. The equilibrium return now converges to $g + y_{\alpha'}^* < g + y_\alpha^*$. Thus, the agents with a lower risk aversion dominates the market, but produce lower equilibrium yield by investing a higher wealth share in the risky asset.

This analysis helps to explain results of the simulations in [7] and [9], and their findings concerning the interplay between risk aversion and memory. We have seen that the risk aversion is mostly related to the capability of agents to invade the market, whereas the memory span influences the stability of the dynamics. These properties hold as long as the investment function on the "EML plot" shifts upward with decrease of the risk aversion. It is easy to see that the investment function, coming from expected utility maximization with power utility has the same general features as mean-variance function used in the examples above. In fact, for a given y and a given perceived variance σ^2, the agents with lower risk aversion invest more, which guarantees the upward shift of the cross-section. As a result, Propositions 3 and 2 can be used. They provide rigorous analytic support of the simulation results of the LLS model.

In [7] the focus is on the role of the memory. The authors show that with a small memory span the log-price dynamics is characterized by crashes and booms. Our analysis shows that this is due to the presence of an unstable equilibrium and to the upper and lower bounds of the investment shares. Furthermore, this equilibrium becomes stable if the memory is high enough. Simulations in [7] confirm this statement; when agents with higher memory are introduced, booms and crashes disappear and price fluctuations become erratic. But as we found, these fluctuations are due to the exogenous noise (coming from the dividend) and not to the endogenous agents' interactions.

In [9] the focus is on the interplay between the length of the memory span and risk aversion. The simulations suggest that the risk aversion is more important than memory in the determination of the dominating agents, providing that the memory is not too short (see the quote in Section 19.1). Our analytic results explains why this is the case. Namely, it is because agents with low risk aversion are able to destabilize the market populated by agents with high risk aversion. However, this "invasion" leads to an ultimate domination only if the invading agents have sufficiently long memory. Otherwise, agents with different risk aversion coefficients will coexist. Notice that this result is new compared to [9] and related works. Thus, our analytic investigation is indeed helpful in understanding the interplay between risk aversion and memory. Another new result concerns the case of agents investing constant fraction of wealth. In [9] the authors claim that such agents always dominate the market and add (p. 571):

> "Hence, the survival of such strategies in real-life markets remains a puzzle within the Levy, Levy and Solomon microscopic simulation framework as it does within the Efficient Market Theory."

Our analysis allows one to understand and also correct this statement. The agents with constant investment fraction are characterized by the horizontal investment functions, for which Proposition 3 guarantees stability independently of L. If these agents are able to invade the market successfully, they will ultimately dominate. However, they cannot invade the market when other agents invest more in their EPE.

Finally, notice that for the case $g > r_f$, which we discuss here, Corollary 1 implies that the economy grows with rate g. All our present and all previous simulations are in accord with this statement. The case $g < r_f$ appears in [6], where the dividend is constant, so that $g = 0$, while the risk-free rate is positive. The resulting price grows with rate r_f, as we can expect from Corollary 1.

19.5 Conclusion

We have performed an analytic investigation of the LLS model and used its results to explain simulations in [6, 7, 8, 9]. We show that the two parameters governing the profitability of the risky and riskless investment opportunities, dividend growth rate and risk-free interest, and determine whether the equity premium can be endogenously generated at equilibrium. The size of equity premium depends on the agents' behavior. We have shown how the stability of the equilibria is related to the memory span that agents use to estimate future returns and their risk aversion. The results are very general and can help understand and extend the findings of previous simulations even when the functional form of the investment function is not known explicitly.

References

[1] Anufriev M (2005) Wealth-Driven Competition in a Speculative Financial Market: Examples with Maximizing Agents. CeNDEF Working Paper 05-17

[2] Anufriev M, Bottazzi G, Pancotto F (2006) Equilibria, Stability and Asymptotic Dominance in a Speculative Market with Heterogeneous Agents. Journal of Economic Dynamics and Control (*forthcoming*)

[3] Anufriev M, Bottazzi G (2006) Price and Wealth Dynamics in a Speculative Market with Generic Procedurally Rational Traders. CeNDEF Working Paper 06-02

[4] Chiarella C, He X (2001) Asset price and wealth dynamics under heterogeneous expectations. Quantitative Finance 1:509-526

[5] Hommes CH (2006) Heterogeneous agents models in economics and finance. In: Judd K, Tesfatsion L (eds) Handbook of Computational Economics II: Agent-Based Computational Economics. Elsevier, North-Holland

[6] Levy M, Levy H and Solomon S (1994) A microscopic model of the stock market: cycles, booms, and crashes. Economics Letters 45, 1:103-111.

[7] Levy M, Levy H (1996) The danger of assuming homogeneous expectations. Financial Analysts Journal May/June:65-70.

[8] Levy M, Levy H and Solomon S (2000) Microscopic simulation of financial markets. Academic Press, London.

[9] Zschischang E, Lux T (2001) Some new results on the Levy, Levy and Solomon microscopic stock market model. Physica A, 291:563-573.

Explaining the Statistical Features of the Spanish Stock Market from the Bottom-Up

José A. Pascual, J. Pajares, and A. López-Paredes

InSiSoc Group. University of Valladolid, Spain

Summary. In this paper, we use an agent based artificial stock market to explore the relations between the heterogeneity of investors behaviour and the aggregated behaviour of financial markets. In particular, we want to recover the main statistical features of the Spanish Stock Market, as the high levels of kurtosis, excess volatility, non normality of prices and returns, unit roots and volatility clustering.

We realise that we cannot catch up most of this features in a market populated only with fundamental investors, so we need to include more heterogeneity in agents behaviour. We include psychological investors who change their risk aversion following the ideas by Kahneman and Tversky (1979) and technical traders who buy or sell depending on crosses of moving averages. The main conclusion is that, in this particular artificial stock market, psychological investors are related to volatility clustering whereas technical trading has more to do with unit roots.

20.1 Introduction

The aim in our research is to explain the behavior of financial markets, and the links between this aggregated macro-behavior and the micro-behavior of investors, filling the gap between the mainstream financial theories and real markets.

Mainstream finance is grounded on strong hypothesis about the rationality of investors; markets are efficient and investors are able to form rational expectations about future value of the relevant variables by means of analyzing the available information. A lot of elegant models have been built under this framework, so we have been able to explain most of the relevant issues in financial markets.

However, real markets exhibit some "stylized facts", which are difficult to explain under this orthodox framework. Among others, excess volatility, non-normality of returns, excess kurtosis, volatility clusters, unit roots, etc.

As suggested in Pajares et al. (2003, 2005), agent based social simulation can help us to explain why these anomalies take place, as we are able to get deeper understanding of the relations between the micro-behavior of individual investors and the macro-behavior of the market. In particular, by means of agent based modeling and simulation, we can built models which include some of the most relevant ideas from behavioral finance (Fama (1998), Shiller(1981)) or the results of the experiments by Kahneman and Tversky (1979) about the real behavior of human beings facing risk.

In this paper, we build an agent based stock market and we introduce different kinds of investors, with different proportions and different trading rules, and we explore the statistical features of the historical series of prices, returns, etc. that emerge in our artificial stock market. We compare theses features with the statistical properties of IBEX-35, the main index of the Spanish stock market. We explore the relations between the proportions of investors exhibiting different behaviors and the statistical features of the aggregated market.

First, we build a model grounded on the artificial stock market by LeBaron et al. (1999), (SFASM), as this model has become a reference in agent based finance. One stock is traded in the market, and it is possible to lend or borrow at the risk free interest rate. Price emerges as a consequence of the bids and offers of shares.

In a first stage, the model is populated with agents who behave in a similar way than the "fundamental investors", in the sense that they process all the available information and form expectations about future prices and dividends. They decide to buy/sell depending on the disagreement of the expectations with real prices.

We have validated this model with LeBaron's, and we have checked that our model produces series with the same statistical features: levels of standard deviation, kurtosis, trading volumes, prices; cross-correlation between squared returns and volume for different lags, etc.

Once we have validated our model, we investigate its financial properties and see that the output of this market is nearly in agreement with the "ideal market" suggested by the financial literature.

But then, we have analyzed the features of the Spanish stock index IBEX-35 and we have noticed some statistical properties which are not in agreement with the output of this agent based model. In particular, se see that the distribution of both prices and returns are not normal, exhibiting fat tails and high kurtosis; there is a unit root in price series; returns are uncorrelated for different lags; and we can see volatility clusters, so the autocorrelations of squared returns are significantly positive even for high lags.

We want to fill the gap between these facts in IBEX-35 and the "ideal market" that emerges from the simple model above. In order to understand the financial underlying concepts, we have broken down the problem into a coupe of steps: first, we introduce psychological investors whose risk aversion changes over time following some of the Kanheman and Tversky ideas; then, we explore the role of technical trading. The main result of our research is that psychological investors are more related with market bubbles whereas technical trading has more to do with the volatility and stationarity of the series of prices.

The exploration of the relations between the micro-behaviour and the statistical properties in financial markets is not new. Morone (2005) shows by means of an experimental laboratory that, both quantity and quality of information are related to fat tails in returns distributions and persistence in volatility. Lux and Marchesi (1999, 2000) explore the role of interactions between traders and the statistical properties of financial markets. To this aim, they study markets populated with fundamental and noise traders.

In this paper, we concentrate on the statistical features of the Spanish Stock Market, and we focus in the role of risk aversion and technical trading. In our work, fundamentalist traders form expectations about future prices and dividends by means of a learning device, psychological investors change their risk aversion over time, and technical dealers use common technical trading rules (moving averages).

The rest of the paper is organized in the following way. In next section (two), we explain the main features of the model. We validate this model with the SFASM. In section three, we present the statistical properties of the IBEX-35, the main index in the Spanish Stock Market. At this point we realize that our initial model, populated only with fundamental investors, cannot catch up some of the statistical features of the Spanish market. We want to fill the gap, so in section four we introduce in the model different proportion of psychological investors whereas in section five we explore the role of technical trading. We finish with the main conclusions of our research.

20.2 The Initial Model with Fundamental Investors

This model has been widely inspired in the SFASM, so we can validate our results with it. A single risky stock is traded in the market and it is also possible to borrow or lend money at the risk free interest rate. Risk free interest rate is constant during the simulation, and there are no transaction costs when lending or borrowing.

For the purpose of this paper, the amount of dividends paid by the risky stock follows an order one auto-regressive model, but we can use any kind of dividend structure. Anyway, dealers do not know, *ex-ante*, the future dividends but they can build models in order to forecast the underlying structure, taking into account all the information available at the trading time. At any period, each investor has to decide, within some budgetary constraints, the number of shares he/she wants to buy or sell and the amount of money to lend or to borrow at the risk free interest rate. Investors send their demands to a specialist who plays the role of a *clearing house*. The specialist does not trade shares at all: he/she just compute the price that clears the market, according to the bids and asks received. We should emphasize here that price is not exogenously fixed, but emerges as a consequence of the interaction between supply and demand of shares. The number of dealers playing in the market (N) is fixed during the simulation, although any investor can stay inactive during long periods of time.

We say that dealers behave as "fundamental investors" because they process all the relevant information in order to form expectations about future prices and dividends and they buy or sell depending of the differences between these expectations and the real prices in the market. Following LeBaron (op.cit)., the demand of shares is computed as:

$$X_{i,t} = [E_{i,t}(p_{t+1} + d_{t+1}) - p_t(1 + r_f)]/\lambda \sigma_{i,t,p+d}^2 \qquad (20.1)$$

where p_t and d_t are prices and dividends in period t, E means expectations, λ is a measure of the risk aversion, and σ^2 is the forecasting variance. So the expectation of futures prices plus dividends next period is compared with the money the agent will get investing the money at the risk free interest rate. Then, this difference is adjusted taking into account the variance of forecasts and the aversion to risk. Agents form expectations by means of a learning mechanism, which is a classifier system. The antecedents of the rules are information concerning the market, whereas the consequents are the parameters the expectations function. The particular issues concerning this learning mechanism can be seen in Pascual (2006).

In order to validate our model with LeBaron's, we have simulated both fast and low learning. In the case of fast learning, agents update their decisions rules by means of the genetic algorithm of the classifier each 250 periods, whereas in the slow learning case, agents need 1000 periods to update the rules. In Table 20.1. we reproduce some statistical numbers for our model and SFASM. The standard deviation of the returns is over 2, and there is a lightly excess kurtosis (it should be cero under the normal distribution). But anyway, the fast learning mode exhibits more excess kurtosis than the slow mode. Something similar happens with the excess return over the risk free interest rate and the trading volume.

Table 20.1. Statistical Data in SFASM and Our Model.

	Fast Learning		Slow Learning	
	SFASM	Ours	SFASM	Ours
Std. Dev.	2.147	2.095	2.135	2.081
Exc. Kurtosis	0.320	0.229	0.072	0.098
P	0.007	0.012	0.036	0.051
Exc. Return	3.062	2.315	2.891	2.183
Trading Volume	0.706	0.434	0.255	0.209

We have also analyzed the autocorrelations of returns, checking that they quickly trend to zero (figure 20.2 left). And, as it happened in LeBaron's model, we also reproduce with our model the fact that the cross correlations of the squared returns with trading volume have peaks between lags −1 and 2. (See figure 20.2 right).

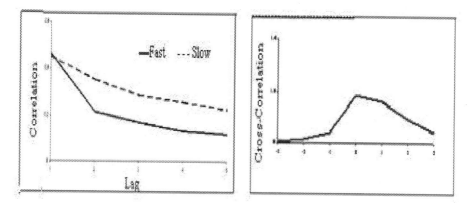

Fig. 20.1. Volume autocorrelations and Correlations of squared returns with volume.

20.3 The Spanish Ibex-35 and the Gap with the "Fundamental Model"

Now we compare the output from our model with a real stock market. The Ibex-35 is the most relevant index in the Spanish market. It is built with the 35 most important companies (in terms trading volume and assets) trading in the "Mercado Continuo". We have analyzed their statistical features in order to check whether they can be reproduced by the "fundamental model". In figure 20.3, we show the evolution of Ibex-35, during the period Jun 01-Jun 05.

We have performed the Jarque-Bera tests and we reject the hypothesis of normality for both prices and returns. Kurtosis of returns is 4.892 (level of 3 for the normal distribution) and the autocorrelations of returns are not significant even for small lags. However, autocorrelations of squared returns are significantly positive even for high lags, which suggest that volatility clusters could be present (see Cont (2001)). (See figure 20.3). We have also performed augmented Dickey-Fuller (ADF) and Phillips-Perron (PP) tests, and we finish that the series of prices has a unit root. We do not detect unit roots in the returns.

Does the "fundamental model" reproduce these facts?. In order to answer this question, we have run simulations in a market populated only with fundamental investors (BFagents).

We have checked that, as it happens for Ibex, the autocorrelations of returns are close to zero even for small lags. Both prices and returns are not

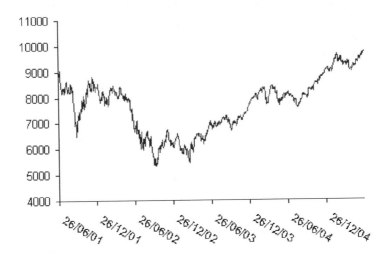

Fig. 20.2. Ibex Evolution

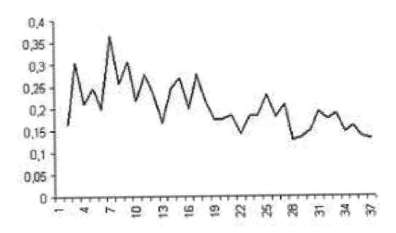

Fig. 20.3. Autocorrelations of squared returns of Ibex 35.

normally distributed. As an example, in table 20.2, we show the main statistical numbers of returns for ten simulations, and the average numbers. We see that the levels of kurtosis are around 3.5, far away from the levels of Ibex 35. We have performed augmented Dickey-Fuller and Phillips-Perron tests, and we do not appreciate unit roots.

In figure 20.3, we show the autocorrelations of squared returns for ten simulations. They are close to zero beyond small lags, so we do not reproduce

Table 20.2. Statistical numbers of returns.

	R0	R1	R2	R3	R4	R5	R6	R7	R8	R9	Mean
Mean	-0.01	-0.02	0.02	0.01	0.01	0.00	-0.01	-0.00	-0.01	0.01	0.00
St. De.	1.909	1.94	2.04	1.82	1.90	2.09	1.84	1.99	2.04	1.98	1.96
Skew.	-0.14	-0.11	0.22	-0.14	0.13	-0.05	0.13	0.03	0.09	-0.19	-0.00
Kurt.	3.34	3.35	3.99	3.33	2.80	3.68	3.84	3.42	3.29	3.27	3.43
J-B	7.85	7.27	48.89	7.83	4.34	19.65	32.21	7.40	4.76	9.27	7.75
Prob.	0.02	0.03	0.00	0.02	0.09	0.00	0.00	0.02	0.09	0.01	0.02

volatility clustering at all; moreover, levels of volatility are low, compared with real markets.

20.4 Filling the Gap. Psychological Investors

Now, we run a market populated with both fundamental and psychological investors. The new investors change their risk aversion over time depending on their previous performance in the market, as suggested by Kanheman and Tversky (1979). Psychological investors also form expectations about futures prices and dividends but, their risk aversion changes depending on the evolution of their wealth, that is, depending on the performance of their previous deals. In particular, we compare the agent's present wealth, with the wealth of the last ten periods.

Tversky and Kahneman (1992) conclude that the importance in risk aversion of a loss is twice the importance of a gain. For this reason, in equation (1), the risk aversion coefficient (λ) can take two values: 0.5 under normal circumstances and 1 whenever the dealer experiments the feeling of loss.

In Table 20.3, we can see the average figures for 10 simulations, for different proportions of fundamental and psychological agents. For instance, 15bf5kt means that the market is populated with 15 fundamental and 5 psychological investors (KT from Kahneman and Tversky).

Kurtosis of returns increases as the number of psychological investors is higher; levels are now closer to the numbers exhibited by Ibex-35. The same is true for the excess volatility. Series are also not normally distributed (probability equal to zero). The autocorrelations of squared returns begin to be significant whenever the proportion of KTagents becomes important (see figure 20.4), which means that volatility clustering appears in markets with high proportion of psychological investors. However, we have not detected unit root in the series of prices, after performing the usual ADF and PP tests.

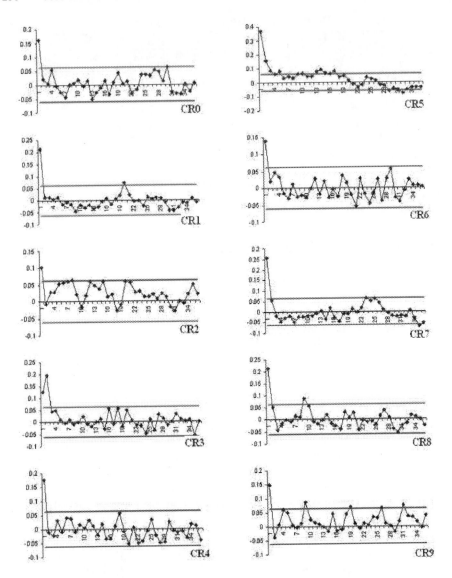

Fig. 20.4. Autocorrelations of squared returns for ten simulations.

20.5 Filling the Gap. Technical Trading

Now, we have built a model with both fundamental and technical dealers. They compute a low order (MA(l)) and a high order moving average (MA(h)) of prices; they buy shares when the MA(l) crosses from down to up to the MA(h) and sell stocks if MA(l) crosses the other one from top to down (see figure 20.5). Technical trading rules like this one are very common in both

Table 20.3. Average figures for 10 simulations

	20BF		15BF5KT		10BF10KT		5BF15KT		20KT	
	P	R	P	R	P	R	P	R	P	R
Mean	81.93	0.00	81.40	0.00	81.35	0.00	79.70	0.00	78.91	0.00
Std. Dev.	7.37	1.96	7.37	1.98	7.52	2.02	8.98	2.26	9.95	2.46
Ex. Var %	81.54		88.53		108.0		169.06		231.95	
Skewness	-0.28	0.00	-0.14	-0.02	-0.28	0.10	-0.65	-0.04	-0.76	-0.09
Kurtosis	4.17	3.43	3.90	3.63	3.78	4.05	4.03	4.63	4.69	6.37
J-B	69.21	7.75	37.37	16.34	38.61	47.33	115.71	110.2	216.15	475.6
Prob	0.00	0.02	0.00	0.00	0.00	0.00	0.00	0.00	0.00	0.00

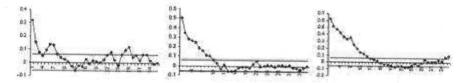

Fig. 20.5. Autocorrelations of squared returns with 15bf5kt, 10bf10kt and 5bf15kt

financial and commodity markets. Even sometimes, the orders are automatically programmed in an Excel sheet which receives on line data form the market.

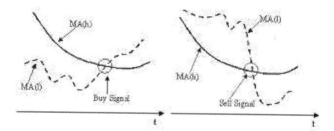

Fig. 20.6. Buying and selling signals in technical trading

In Table 20.4, we show average figures for different proportions of technical traders (TFagents). When the number of them increases, both excess volatility and prices increases significantly. Normality of both individual series of prices and returns is rejected, although the probability computed with mean S and mean K is different to 0 in some cases. Kurtosis also evolves to the levels exhibited by Ibex-35, and the higher the proportion of technical traders, the greater the excess of kurtosis over normal distribution. Although we also have got evidence of volatility clusters, this evidence is not as strong as it was in the previous case and in the real Spanish market, as the number of lags with significant squared autocorrelations is much lower than in the previous case.

Table 20.4. Average figures for 10 simulations

	20BF		15BF5TF		10BF10TF		5BF15TF	
	P	R	P	R	P	R	P	R
Mean	81.93	0.00	99.03	-0.01	105.97	-0.02	107.23	-0.01
Std. Dev.	7.37	1.96	9.36	1.61	18.58	1.46	20.99	1.30
Ex. Var %	81.54		215.86		1207.22		1398.98	
Skewness	-0.28	0.00	0.25	0.16	0.01	0.13	0.10	0.35
Kurtosis	4.17	3.43	3.86	4.09	2.71	4.04	2.92	5.54
J-B	69.21	7.75	41.11	53.96	3.51	47.57	1.88	289.24
Prob	0.00	0.02	0.00	0.00	0.17	0.00	0.39	0.00

However, we see that technical trading could be related with unit roots. In Table 20.5, we show the results of the ADF and PP test for ten simulations, with different proportions of technical dealers. The evidence of unit roots is even higher whenever the proportion of technical traders increases.

Table 20.5. ADF and PP Tests for different simulations

	20bf		15bf5tf		10bf10tf		5bf15tf		Critical
Simulation	ADF	PP	ADF	PP	ADF	PP	ADF	PP	Value
0	-4.12	-4.07	-5.13	-4.85	-2.66	-2.19	-2.94	-2.10	1% -3.44
1	-5.37	-5.27	-3.60	-3.21	-4.30	-3.79	-2.53	-1.75	5% -2.86
2	-3.75	-3.54	-4.77	-4.31	-1.42	-1.40	-2.42	-1.84	10% -2.57
3	-4.86	-4.68	-4.36	-4.14	-1.87	-1.67	-2.37	-1.84	
4	-5.09	-4.77	-2.97	-2.69	-2.08	-1.89	-2.10	-1.22	
5	-4.97	-4.46	-4.26	-3.94	-3.49	-2.92	-2.19	-1.25	
6	-4.36	-4.25	-4.34	-3.80	-3.00	-2.47	-3.04	-2.61	
7	-4.16	-3.87	-4.48	-3.83	-1.79	-1.48	-2.72	-2.07	
8	-4.58	-4.18	-3.94	-3.82	-2.21	-2.02	-2.74	-1.81	
9	-4.78	-4.16	-4.24	-4.07	-2.43	-2.01	-2.73	-1.72	

20.6 Conclusions

We have built an artificial stock market which reproduces the main statistical features of the Ibex-35, the main Spanish Stock Market index. In order to achieve this purpose, we have run different markets with different proportions of fundamental, psychological and technical investors. The first important conclusion is that all kinds of investors are necessary if we want to reproduce the index, as the presence of fundamental dealers alone do not explain some statistical properties.

Non normality of prices and returns and excess kurtosis and high volatility is reasonably well explained by means of technical and psychological investors. We have learnt from simulations that psychological trading help us to understand the emergence of volatility clustering, whereas technical trading has more to do with higher levels of kurtosis and the existence of unit roots in returns.

Of course, the model could be improved, in order to include more investor behaviors, more stocks, etc., but the purpose of this initial work was to explore the relations between heterogeneity in the market and its statistical properties, so that we could reproduce the statistical features of the Spanish Stock Market. Our work wants to be an example of "Generative Science", in the sense by Epstein and Axtell, (1996, 1997); Axelrod (1997).

Acknowledgments

We want to thank to Professor Cesareo Hernandez, for his important ideas and his contributions to this work. He encouraged us to explore the relations between market heterogeneity and aggregated performance of the stock markets.

This research has been partially funded by the Spanish Ministry of Education and Culture, Research and Development Program, grant Project DPI2005-05676.

References

[1] Cont R. (2001) Empirical properties of asset returns: stylised facts and statistical issues. *Quantitative Finance*, Vol 1, pp 223-236.
[2] Axelrofd (1997) "Advancing the Art of Simulation in the Social Sciences", *Complexity* Vol. 3, pp. 193–199.
[3] Epstein, J. M., and Axtell, R. (1996) *Growing Artificial Societies: Social Science From the Bottom Up.* Cambridge, MA: MIT Press.
[4] Epstein, J.M.; Axtell, R. (1997). "Artificial Societies and Generative Social Science". *Artificial Life and Robotics* Vol. 1, pp. 33–34.
[5] Fama, E. F., (1998) "Market efficiency, long-term returns, and behavioral finance", *Journal of Financial Economics*, Vol. 49. pp. 283-306.
[6] Kahneman D., and Tversky A (1979) Prospect theory: An analysis of decisions under risk". *Econometrica*, 47, pp 313 327.
[7] LeBaron B., Arthur W.B. and Palmer R. (1999) Time series properties of an artificial stock market. *Journal of Economic Dynamics and Control*, vol 23, pp 1487-1516.

[8] Lux, T. and Marchesi, M. (1999). "Scaling and criticality in a stochastic multi-agent model of a financial market". *Nature*, 397:498—500.

[9] Lux, T. and Marchesi, M. (2000). "Volatility Clustering in financial markets: a micro-simulation of interacting agents". *International Journal of Applied Finance*, 3, pp: 675-702.

[10] Morone, A. 2005. "Financial Market in the Laboratory, an Experimental Analysis of some Stylized Facts," *Discussion Papers on Strategic Interaction 2005-27*, Max Planck Institute of Economics, Strategic Interaction Group

[11] Pajares J., Pascual J.A., Hernández C. and López-Paredes A. (2003) A behavioural, evolutionary and generative approach for modelling financial markets. 1st *Conference of the European Social Simulation Association (ESSA)*. Groningen. The Netherlands.

[12] Pajares J., Pascual J.A., Hernández C. and López-Paredes A. (2005). The role of risk aversion and technical trading in the behaviour of financial markets 3rd *Conference of the European Social Simulation Association (ESSA)*. Koblenz. Germany. Sept 2005.

[13] Pascual J. A. (2006). Modelado Multiagente de Mercados Financieros: Un Enfoque basado en el Comportamiento Individual de los Inversores. Tesis Doctoral. Departamento de Organización de Empresas y C.I.M. ETS de Ingenieros Industriales. Universidad del Valladolid.

[14] Shiller, R. J. (1981) "Do Stock Prices Move Too Much to be Justified by Subsequent Changes in Dividends?" *The American Economic Review*, Vol. 1, Issue 3, pp. 421-436

[15] Tversky A and Kahneman D (1992), "Advances in Prospect Theory: Cumulative Representation of Uncertainty". *Journal of Risk and Uncertainty*, Vol. 5 (4) pp. 297-323.

Lecture Notes in Economics and Mathematical Systems

For information about Vols. 1–489
please contact your bookseller or Springer-Verlag

Vol. 533: G. Dudek, Collaborative Planning in Supply Chains. X, 234 pages. 2004.

Vol. 534: M. Runkel, Environmental and Resource Policy for Consumer Durables. X, 197 pages. 2004.

Vol. 535: X. Gandibleux, M. Sevaux, K. Sörensen, V. T'kindt (Eds.), Metaheuristics for Multiobjective Optimisation. IX, 249 pages. 2004.

Vol. 536: R. Brüggemann, Model Reduction Methods for Vector Autoregressive Processes. X, 218 pages. 2004.

Vol. 537: A. Esser, Pricing in (In)Complete Markets. XI, 122 pages, 2004.

Vol. 538: S. Kokot, The Econometrics of Sequential Trade Models. XI, 193 pages. 2004.

Vol. 539: N. Hautsch, Modelling Irregularly Spaced Financial Data. XII, 291 pages. 2004.

Vol. 540: H. Kraft, Optimal Portfolios with Stochastic Interest Rates and Defaultable Assets. X, 173 pages. 2004.

Vol. 541: G.-y. Chen, X. Huang, X. Yang, Vector Optimization. X, 306 pages. 2005.

Vol. 542: J. Lingens, Union Wage Bargaining and Economic Growth. XIII, 199 pages. 2004.

Vol. 543: C. Benkert, Default Risk in Bond and Credit Derivatives Markets. IX, 135 pages. 2004.

Vol. 544: B. Fleischmann, A. Klose, Distribution Logistics. X, 284 pages. 2004.

Vol. 545: R. Hafner, Stochastic Implied Volatility. XI, 229 pages. 2004.

Vol. 546: D. Quadt, Lot-Sizing and Scheduling for Flexible Flow Lines. XVIII, 227 pages. 2004.

Vol. 547: M. Wildi, Signal Extraction. XI, 279 pages. 2005.

Vol. 548: D. Kuhn, Generalized Bounds for Convex Multistage Stochastic Programs. XI, 190 pages. 2005.

Vol. 549: G. N. Krieg, Kanban-Controlled Manufacturing Systems. IX, 236 pages. 2005.

Vol. 550: T. Lux, S. Reitz, E. Samanidou, Nonlinear Dynamics and Heterogeneous Interacting Agents. XIII, 327 pages. 2005.

Vol. 551: J. Leskow, M. Puchet Anyul, L. F. Punzo, New Tools of Economic Dynamics. XIX, 392 pages. 2005.

Vol. 552: C. Suerie, Time Continuity in Discrete Time Models. XVIII, 229 pages. 2005.

Vol. 553: B. Mönch, Strategic Trading in Illiquid Markets. XIII, 116 pages. 2005.

Vol. 554: R. Foellmi, Consumption Structure and Macroeconomics. IX, 152 pages. 2005.

Vol. 555: J. Wenzelburger, Learning in Economic Systems with Expectations Feedback (planned) 2005.

Vol. 556: R. Branzei, D. Dimitrov, S. Tijs, Models in Cooperative Game Theory. VIII, 135 pages. 2005.

Vol. 557: S. Barbaro, Equity and Efficiency Considerations of Public Higer Education. XII, 128 pages. 2005.

Vol. 558: M. Faliva, M. G. Zoia, Topics in Dynamic Model Analysis. X, 144 pages. 2005.

Vol. 559: M. Schulmerich, Real Options Valuation. XVI, 357 pages. 2005.

Vol. 560: A. von Schemde, Index and Stability in Bimatrix Games. X, 151 pages. 2005.

Vol. 561: H. Bobzin, Principles of Network Economics. XX, 390 pages. 2006.

Vol. 562: T. Langenberg, Standardization and Expectations. IX, 132 pages. 2006.

Vol. 563: A. Seeger (Ed.), Recent Advances in Optimization. XI, 455 pages. 2006.

Vol. 564: P. Mathieu, B. Beaufils, O. Brandouy (Eds.), Artificial Economics. XIII, 237 pages. 2005.

Vol. 565: W. Lemke, Term Structure Modeling and Estimation in a State Space Framework. IX, 224 pages. 2006.

Vol. 566: M. Genser, A Structural Framework for the Pricing of Corporate Securities. XIX, 176 pages. 2006.

Vol. 567: A. Namatame, T. Kaizouji, Y. Aruga (Eds.), The Complex Networks of Economic Interactions. XI, 343 pages. 2006.

Vol. 568: M. Caliendo, Microeconometric Evaluation of Labour Market Policies. XVII, 258 pages. 2006.

Vol. 569: L. Neubecker, Strategic Competition in Oligopolies with Fluctuating Demand. IX, 233 pages. 2006.

Vol. 570: J. Woo, The Political Economy of Fiscal Policy. X, 169 pages. 2006.

Vol. 571: T. Herwig, Market-Conform Valuation of Options. VIII, 104 pages. 2006.

Vol. 572: M. F. Jäkel, Pensionomics. XII, 316 pages. 2006.

Vol. 573: J. Emami Namini, International Trade and Multinational Activity, X, 159 pages, 2006.

Vol. 574: R. Kleber, Dynamic Inventory Management in Reverse Logisnes, XII, 181 pages, 2006.

Vol. 575: R. Hellermann, Capacity Options for Revenue Management, XV, 199 pages, 2006.

Vol. 576: J. Zajac, Economics Dynamics, Information and Equilibnum, X, 284 pages, 2006.

Vol. 577: K. Rudolph, Bargaining Power Effects in Financial Contracting, XVIII, 330 pages, 2006.

Vol. 578: J. Kühn, Optimal Risk-Return Trade-Offs of Commercial Banks, IX, 149 pages, 2006.

Vol. 579: D. Sondermann, Introduction to Stochastic Calculus for Finance, X, 136 pages, 2006.

Vol. 580: S. Seifert, Posted Price Offers in Internet Auction Markets, IX, 186 pages, 2006.

Vol. 581: K. Marti; Y. Ermoliev; M. Makowsk; G. Pflug (Eds.), Coping with Uncertainty, XIII, 330 pages, 2006 (planned).

Vol. 582: J. Andritzky, Sovereign Default Risk Valuation (planned).

Vol. 583: I. Konnov, Generalized Convexity and Related Topics (planned).

Vol. 584: C. Bruun, Advances in Artificial Economics, XVI, 294 pages, 2006.